Community Psychology

Series Editors

Mohamed Seedat
Institute for Social and Health Sciences, University of South Africa, Johannesburg, South Africa
South African Medical Research Council-University of South Africa Masculinity and Health Research Unit, Cape Town, South Africa

Shahnaaz Suffla
Institute for Social and Health Sciences, University of South Africa, Johannesburg, South Africa
South African Medical Research Council-University of South Africa Masculinity and Health Research Unit, Cape Town, South Africa

The Community Psychology book series is envisaged as a space to review the established assumptions and knowledge economy underlying community psychology, and encourage writings that recognize the plurality of people and the many geographical, psychological and sociological locations that they occupy. The book series will enable contributors to stimulate thought that questions that which is constructed as critical knowledges, community psychology, and the meanings of liberation and community. Contributions to the book series draw attention to the applications of community psychology in the Global South and the Global North as they relate to such issues as violence, socio-economic inequality, racism, gender, migration, dispossession, climate change, and disease outbreaks. In doing so, it centers community psychology as focused on the well-being of collectives, and dealing with such focal issues as deploying psychology to support social justice, the relevance and appropriateness of its internal logic, and methods that deal with the range of psychological, social, cultural, economic, political, environmental, epistemic, and local and global influences that bear on the quality of life of individuals, communities and society. The book series concentrates thus on the following three key areas of focus: 1) decoloniality, power and epistemic justice, 2) knowledge production, contestation and community psychology, and 3) community psychology in context. The series is of vital and immediate relevance to researchers, practitioners, faculty and students from the intervention sciences, including anthropology, sociology, public health, development studies, social work and urban studies.

More information about this series at http://www.springer.com/series/15965

Garth Stevens • Christopher C. Sonn
Editors

Decoloniality and Epistemic Justice in Contemporary Community Psychology

Springer

Editors
Garth Stevens
Department of Psychology
University of Witwatersrand
Johannesburg, Gauteng, South Africa

Christopher C. Sonn
Institute of Health and Sport
Victoria University
Melbourne, VIC, Australia

ISSN 2523-7241 ISSN 2523-725X (electronic)
Community Psychology
ISBN 978-3-030-72219-7 ISBN 978-3-030-72220-3 (eBook)
https://doi.org/10.1007/978-3-030-72220-3

This Springer imprint is published by the registered company Springer Nature Switzerland AG
The registered company address is: Gewerbestrasse 11, 6330 Cham, Switzerland

Series Foreword
The Community Psychology Book Series: A Dialogical Decolonizing Space

The Springer Community Psychology Book Series holds an ambitious vision that approaches community psychology as a site of knowledge and knowledge-making. The Series is imagined as a dialogical space for critical and situated knowledges and liberatory praxes. Through interrogations of decolonizing psychologies' obligations in the era of the modern knowledge economy and the embrace of alternative, disruptive, and new imaginings, the Book Series intends to elaborate on collective critical-liberatory projects in community psychology and cognate areas of scholarship. The Series is however alert to the limitations imposed on creative thought and practices by hierarchical and homogenizing influences in and outside of the academy.

The Series encourages contributions that focus on community psychology as knowledge and the contestation for representation and authority. We invite contributors to examine the politics and archeology of knowledge. Likewise, the Series draws together contributions that review how particular concepts and theories have gained ascendency in community psychology, and that offer insights into the benefits and limitations of creative methodologies applied in research, intervention, and analysis.

The Series encourages work that centrally engages with "epistemicide"—the deracination of other-than Western modes of knowing and knowledge systems—and its manifestations in scholarship on community, community-making, community resistance, and the formations of social arrangements that seek to overturn racism, racialization, heteropatriarchy, classism, and other forms of oppressive social relations. The Series is aligned with and situated within the larger body of praxes of the South, borne out of struggles for self-determination, epistemic independence and epistemic agency, and visions and imaginations of radical humanism. The Series thus seeks to animate conversations about what it means to create and live in human formations that challenge "race" and racism, gendered and patriarchal arrangements, inequitable economic and material arrangements, and problematic notions of sex and sexuality, as well as a range of other exclusionary-isms.

This book, *Decoloniality and Epistemic Justice in Contemporary Community Psychology*, the third in the Series and co-edited and led by Garth Stevens and Christopher C. Sonn, presents epistemologies from the Global South and the South in the Global North, and their interpretations of community, liberation, and collective psychologies. Contributors, articulating multiple critical perspectives from different regions of the world, probe the manner in which decolonial thought has obtained presence in various current enactments and writings of community psychology. Contributors problematize the inordinate influences of Euroamerican thought and illustrate how epistemic alternatives may open and free up theoretical spaces to reconsider community psychology as liberatory knowledge and practice. This edited text is a seminal theoretical intervention that helps reshape the identities and representational forms of knowledge on community. The anthology of scholarly and activist essays elaborates and supports the collective critical project in community psychology and cognate disciplines. *Decoloniality and Epistemic Justice in Contemporary Community Psychology* is an impressive and incisive offering that will undoubtedly inspire its readership.

Mohamed Seedat
Shahnaaz Suffla

Prologue
Decolonial Psychology as a Counter-Catastrophic Science

If, far from being the undisputable source of progress and development, globalized Western modernity is a generator of catastrophe, then any decolonial turn in psychology must by necessity lead to the production of counter-catastrophic psychologies. At the same time, since professional and academic psychologies are themselves part of a hegemonic order of knowledge that is entangled with coloniality, then any counter-catastrophic psychology needs to also become, in one dimension or another, a form of counter-psychology.

The catastrophe of modernity/coloniality is not just one catastrophe among others. Because of its scope, it could be better understood as a meta-catastrophe with demographic, environmental, and metaphysical dimensions. The catastrophe of modernity/coloniality is also unique in its gravity and tragedy: it is a catastrophe that indicates not only a turn to violence, but the very normalization and naturalization of war.

In the catastrophe of modernity/coloniality, lands are taken from people who are not considered people while the dispossessed, murdered, raped, and tortured non-people and everything related to them—their creations, habitats, and even self-perceptions—become the targets of endless wars. Counter-catastrophic decolonial psychologies are therefore psychologies in times of war, which points to one of psychology's main challenges since psychology, as both a science and clinical practice, does not have the best record when it comes to addressing war.

Consider that it was only a few years ago that the American Psychological Association (APA) was forced to respond to the revelations of the key role of psychology in the design of "enhanced interrogation" techniques by the Central Intelligence Agency in the United States' so-called "war on terror." The U.S. Senate Intelligence Committee found that the CIA apparently paid millions of dollars to two psychologists who "reverse-engineered military training techniques" and who "developed theories of interrogation based on 'learned helplessness'" (Dilanian, 2014; Fish, 2014). While the leadership of the APA expressed outrage, there were also reports that the APA itself had given "greater professional cover for psychologists who had been helping to monitor and oversee harsh interrogations" (p. 194) shortly after the start of the "war on terror" in 2001.

The relevance of the connection between psychology and "enhanced interrogations," that is, torture, for a decolonial turn in psychology becomes especially clear when one considers that, in modernity/coloniality, torture is not an extraordinary phenomenon that represents a deviation from an ethical rule, but rather a crucial practice and logic in the unfolding of coloniality. Torture is part of what Frantz Fanon referred to as a "polydimensional method" of dehumanization, which includes, "[e]xploitation, tortures, raids, racism, collective liquidations, [and] rational oppression." Fanon wrote that these practices "take turns at different levels in order literally to make of the native an object in the hands of the occupying nation." As Fanon explains: "This object man, without means of existing, without a *raison d'être*, is broken in the very depth of his substance" (Fanon, 1967, p. 35). This brokenness of the mind, body, and soul indicates a primary site of catastrophe and opens up the horizon for counter-catastrophic psychology.

While Fanon observed that techniques of dehumanization keep evolving, and that vulgar racism and widespread bodily torture may not always appear prominently in a given society, for him it was clear that "[t]orture is inherent in the whole colonialist configuration" (p. 64). If modernity/coloniality is a paradigm of endless war, then coloniality breeds not only death and early death, but also multiple forms of torture. Fanon observed the workings of torture closely in the Algerian war for independence from France. In 1960, at the height of the war for liberation he wrote:

> A defenceless people is mown down by machine guns and armoured cars, women and children are machine-gunned from the air, policemen indulge in torture—such has been the natural order of existence in Algeria for over the past six years (Fanon, 2018, pp. 665–666).

Coloniality, like the Algerian war that "plunged Africa into mourning," is a "tragic and terrible" catastrophe (Fanon, 1967, pp. 665–666) that turns torture into "a way of life" (Fanon, 1967, p. 66). Therefore, torture by the French in Algeria was not "an accident…an error, or a fault" (p. 66).

Fanon concluded that "[c]olonialism cannot be understood without the possibility of torturing, of violating, or of massacring…. Torture is an expression and a means of the occupant-occupied relationship" (1967, p. 66). This point suggests that targeted massive deaths, rapes, and tortures define normal practices within the symbolic and practical universe of coloniality. This means that massacre, rape, and torture always appear on the horizon of racial and modern/colonial dynamics. Massacre, rape, and torture inspire interpretations of the world, of selfhood, and of others and can find expression in forms that hide their sinister character. They can also appear in their most concrete and brutal expressions at any given time. Names such as those of Eric Garner, Sandra Bland, George Floyd, Breonna Taylor, and Ahmaud Arbery come to mind here.

For massacre, rape, and torture to become a way of life, reason, ethics, and politics have to undergo catastrophic mutations, which in turn create a new apparent order and set of norms. In this context, the massacre, rape, and torture of racialized communities and their descendants are taken as non-problematic or as secondary problems. There is also an overwhelming suspicion that racialized communities themselves are responsible for the problems that they face and that it is them who most clearly represent a problem or challenge to the order. Then, when it is not possible to hide the problems or make the primary victims appear as culpable, another presupposition becomes explicit: that, to the extent that there are actions that anticipate or cultivate massacre, rape, and torture, these are only exceptional cases to be treated individually.

There are three catastrophic operations here: deny the problem, make the victim appear as the problem, and only admit evidence on the basis that it is exceptional and that it can be addressed by focusing on individual behavior, not a collective, a culture, an institution, or a system. These three basic operations play a central role in the production, reproduction, normalization, and naturalization of demographic, environmental, and metaphysical catastrophe. Fanon found them multiple times, both in politics and knowledge production, including in the profession of psychology. For example, as a Director of a psychiatric hospital Fanon was supposed to attend to the immediate needs of torturers and some tortured without directly challenging the legitimacy of torture itself. Fanon found that French doctors and intellectuals were accustomed to rationalize torture when it was used against Algerians, and to consider torture an exception of French behavior and French morals. French psychiatrists and psychologists were also used to approach Algerians with problems as problem people, to the point of behaving as if there was a "North-African syndrome" (Fanon, 1967, p. 3).

Fanon did not have to wait to be in Algeria to discover the catastrophic character of academic knowledge or disciplinary psychology. Fanon puts psychology in the dock and subjects it to critical analysis in the fourth chapter of *Black Skin, White Masks*, which was written when he was a student of psychiatry in France. The culprit in this chapter is Octave Mannoni's *Prospero and Caliban: The Psychology of Colonization*.

That the work of a white French psychoanalyst takes center stage in the fourth chapter of *Black Skin, White Masks* is highly relevant for the book's main argument. Consider that the previous chapters focus on the use of language and love as masks. The main characters in Fanon's play/analysis in the first three chapters are Black men and women who exhibit different forms of anti-Black attitudes. In chapter four, a white psychologist appears as the representative of the established science, decency, and morals. Fanon does not think that either Mannoni's race or the country of his origin makes him unfit to study colonialism. A proper non-catastrophic education and treatment could have effectively challenged at least some problematic presuppositions. Fanon found that this was not the case as he concluded that "although [Mannoni] has devoted 225 pages to the study of the colonial situation, [he] has not grasped the true coordinates" (p. 65).

A question emerges here: what would it take for psychology to grasp the true coordinates of coloniality? There is a large array of responses to this question in this volume, which can be put in conversation with Fanon's analysis, his view of combatting torture as a way of life, and his critique of Mannoni. Fanon's critique of Mannoni could be summarized in a nutshell: not only did Mannoni fail to grasp the true coordinates of colonialism with his psychoanalysis, but his work naturalized the relationship between colonizer and colonized. After exploring the failures of Black colonial subjects to escape the catastrophic practice of seeking to wear "white masks" in the first three chapters of *Black Skin, White Masks*—and is this not but a form of self-torture?—chapter four focuses on Mannoni to illustrate a catastrophic abdication of the intellect and of morals, as well as an active effort to naturalize colonialism and, ultimately we could add with the Fanon of *Toward the African Revolution*, *Wretched of the Earth*, and of his psychiatric and political writings, torture. Yet, this abdication and activity are "normal" in a context marked by coloniality, which means that the coloniality of knowledge both shields and promotes massacre, rape, and torture as ways of life.

In the face of the systemic and systematic character of racism, Mannoni saw pathologies that naturalized the relationship between colonizers and colonized and between whites and blacks. Mannoni not only followed this insight with regard to the people of Madagascar, where he lived for more than 20 years, but also in relation to white and Black South Africans. Fanon's response to Mannoni is worth citing at length. Fanon asks, "What is South Africa?," to which he responds as follows:

> A powder keg where 2,530,300 whites cudgel and impound 13 million Blacks. If these poor Whites hate the Blacks it's not, as Monseiur Mannoni implies, because 'racialism is the work of petty officials, small traders and colonials, who have toiled much without great success.' No, it's because the structure of South Africa is a racist structure… (Fanon, 2008, p. 68).

A racist structure is one where massacre, rape, and torture become, as we have seen, ways of life.

Fanon's analysis indicates that it might not just be Algeria, Madagascar, or South Africa, but the entirety of Africa which could be considered a space profoundly marked by demographic, environmental, and metaphysical catastrophe. The Caribbean and the Americas also appeared to Fanon in a similar light. By making these connections Fanon's work helps to critically engage and undo another catastrophic element in knowledge production: the segregationist and neo-segregationist practices that reify national borders or continental differences. Fanon teaches that while no communities or places are exactly alike, that life under catastrophe, facing the challenge of massacre, rape, and torture as ways of life, is part of what it means to inhabit the South, including the Souths in the North and the Souths of the South. It should therefore not be a surprise that decolonial turns in knowledge production have become prominent activities in the South.

But how exactly to achieve a decolonial turn in psychology? How to outline the principles of a combative counter-catastrophic psychology that is also a form of counter-psychology? One lesson that Fanon's work teaches us is that no decolonial turn in psychology or any other area is possible if one is not willing to leave those areas behind. For Fanon the decolonial project had priority over any existing discipline, which means that for a decolonial turn to take place in psychology, a psychologist must be prepared not only to engage in forms of counter-psychology but also to abandon psychology as a discipline and profession altogether. Abandoning psychology, or any discipline or profession for that matter, is not only a metaphor, but also a concrete possibility. Fanon made as much clear in his resignation from the position as Director of the Psychiatric Hospital at Blida-Joinville before openly joining the Algerian Liberation Front (see, the "Letter to the Resident Minister" in Fanon, 1967, pp. 52–54, and in Fanon, 2018, pp. 433–436).

While Fanon abandoned psychology as a profession, or at least resigned from the system of recognition and authority in the field, he never ceased to care for health and healing. Fanon perceived his resignation from his position in psychiatry as consistent with his view of a medical or healing imperative that was at the center of a counter-catastrophic psychiatric practice. Consider how Fanon explains his decision for his resignation: "If psychiatry is the medical technique that sets out to enable individuals no longer to be foreign to their environment, I owe it to myself to state that the Arab, permanently alienated in his own country, lives in a state of absolute depersonalization" (Fanon, 2018, p. 434). Since Fanon found that the status of Algeria was one of "a systematic dehumanization" and that its "extant social structure…stood opposed to any attempt to put the individual back in his or her place" (p. 434), Fanon resigned from his professional position and joined the revolution. He gave priority to a people and their struggle, not to a discipline or a profession. In a context where the Algerian people were in revolt, he could not but join the revolt himself.

Fanon's earlier psychological and philosophical text, *Black Skin, White Masks*, also becomes relevant to understand Fanon's perception of the medical and decolonial imperative that prioritizes people over disciplines, professions, or existing structures. I have already stated that from a Fanonian point of view, there is no decolonial turn in psychology without the willingness to resign from established positions of authority and legitimacy as well as from hegemonic criteria of merit and distinction. Fanon also makes clear that this resignation is grounded on a commitment with a medical or healing imperative that is part of a deep concern with Black and colonized people in struggle. Elsewhere (see Maldonado-Torres, 2008), in conversation with the work of the Chicana feminist Chela Sandoval, I have referred to this deep concern with Black and colonized people as decolonial love, and I have characterized decolonial love as a central component of the decolonial attitude. I have also argued that for Fanon, attitude has priority over methods and disciplines, and that, therefore, decolonial love and the decolonial attitude are the basis of decolonial turns that generate new questions and new forms of knowledge (Maldonado-Torres, 2017). This means that a decolonial turn in psychology not only depends, as we have seen, on a willingness to resign from a profession, discipline, or system of professional recognition, but also that this very willingness to resign is itself part of a more encompassing decolonial attitude and decolonial love.

It is indeed decolonial love that transpires in Fanon's view of the medical imperative and why the medical imperative is also a decolonial one. In short, Fanon resigns his position as Director of a Psychiatric Hospital out of love, a love that leads him to discover the social and structural dimension of alienation and depersonalization, and a love that leads him to abandon professional recognition for the sake of collaboration with others seeking to create the "world of *you*" (Fanon, 2008, p. 206). It is not necessary to look too far to find Fanon's view of decolonial love and the decolonial attitude. Consider the fourth chapter of *Black Skin, White Masks* once more. Right below the chapter's title and before we start to read Fanon's analysis of Mannoni's colonial psychology, there is an epigraph that encapsulates Fanon's view of a fundamentally different approach to knowledge production and to medical practice. It is as if Fanon was making explicit the basis of his own approach to psychology; a starting point that made all the more evident the catastrophic dimension of Mannoni's failure to understand colonialism and of modern/colonial psychology.

The epigraph in chapter four reads: "There is not in the world one single poor lynched bastard, one poor tortured man, in whom I am not also murdered and humiliated" (Fanon, 2008, p. 64). This sentence appears in a play written by the Afro-Caribbean writer, poet, and intellectual Aimé Césaire. That Fanon considered Césaire a crucial figure for his own thinking notwithstanding Fanon's critique of certain dimensions of the "Negritude" poetic movement, or his disillusionment with Césaire's politics later on is evident in that Fanon also selects a statement from Césaire as the epigraph for the introduction of *Black Skin, White Masks*. This other epigraph, taken from Césaire's anti- and decolonial manifesto *Discourse on Colonialism*, reads "I am talking about millions of men whom they have knowingly instilled with fear and a complex of inferiority, whom they have infused with despair and trained to tremble, to kneel and behave like flunkeys" (quoted in Fanon, 2008, p. xi).

The "Introduction" and the fourth chapter of *Black Skin, White Masks* are closely connected in that, together, they offer the clearest indications of Fanon's critique of established psychological practice as well as a basic outline of his own preferred approach to human problems. In the introduction, Fanon challenges the dominance of ontogeny and phylogeny in psychoanalysis and he adds the area of sociogeny. He also establishes the priority of attitude over methods and asks the reader to "leave methods to the botanists and mathematicians" (Fanon, 2008, p. 15; Maldonado-Torres, 2017). In turn, chapter four engages in a critique of a psychological approach to colonization, that of Mannoni, showing the catastrophic results that ensue when ignoring the meditations on psychology and racism that Fanon offers in the introduction. The fourth chapter is also the only chapter with an epigraph; the other two sections with epigraphs are the introduction and the conclusion. My argument here is that the epigraphs that appear in the interrelated introduction and fourth chapter are also connected to each other and that, together, they point to the basis of Fanon's own critical engagement with psychology and of his own counter-psychological practice. That both epigraphs come from an Afro-Caribbean poet, playwright, and essayist—Césaire—and not from psychologists is already a clear indication of Fanon's counter-psychological tendencies.

Césaire's voice at the start of the introduction and the fourth chapter is key to point to the kind of attitude that Fanon considers fundamental to engage in a decolonial turn in psychology. The "poor lynched bastard" and the "poor tortured man" that are mentioned in one epigraph are part of the millions of people "whom they have knowingly instilled with fear and a complex of inferiority, whom they have infused with despair and trained to tremble" who are referenced in the other epigraph. Reading the two epigraphs side by side suggests indissoluble ties between torture, on the one hand, and the problems of colonial alienation and depersonalization, on the other. This connection also indicates that if colonial alienation and depersonalization involve the interiorization of the point of view of the colonizer, the master, and/or the white in one's mind, then, colonial alienation and depersonalization can also be interpreted as forms of self-torture. In the modern/colonial world, torture is thus clearly, at the objective as well as at the subjective and most intimate levels of subjectivity, a way of life. Decoloniality should therefore be a practice, not only about dispossession—which, if taken at face value would risk to make possession or repossession the goal—but also against endless war and torture.

Another crucial idea that emerges by reading the epigraphs of the introduction and the fourth chapter of Black Skin, White Masks in connection is that the mechanisms that lead to lynching and torture are the same ones that generate fear and the development of feelings of inferiority. In response to this state of affairs, Fanon can only think of a healthy "I" when this "I" is re-constituted as part of a collective effort to end the world as we know it (see Fanon, 2008, p. 191; see also Césaire, 2017, p. 107). By ending the world as we know it, Fanon means the creation of a "world of *you*" (Fanon, 2008, p. 206) which, in the "tragic and terrible" catastrophic conditions that we live in, can only emerge when we are guided by the idea that "[t]here is not in the world one single poor lynched bastard, one poor tortured man, in whom I am not also murdered and humiliated" (quoted in Fanon, 2008, p. 64). That this Césairean declaration of decolonial love becomes the cornerstone of Fanon's own decolonial turn in psychology becomes clear when he states clearly and explicitly his "true wish," which can be taken as the principle of his counter-psychological practice: "My true wish is to get my brother, black or white, to shake off the dust from that lamentable livery built up over centuries of incomprehension" (Fanon, 2008, p. xvi). For this task the counter-psychologist has to, first, undergo a process of unlearning from the discipline so as to understand the gravity of the problem and discover the face of another human behind the masks. The counter-psychologist needs to combat the specter of Mannoni with the voices and images of the poets, essayists, visual artists, playwrights, movement leaders, community organizers, and performers who reveal the catastrophic dimensions of our world—how massacre, rape, and torture continue to be a way of life—and who conjure and enact other ways of knowing, understanding, feeling, and relating with each other. Here lies the promise of a decolonial turn in psychology and the creation of a psychological practice that always prioritizes the cultivation of relationships with collectives of Black, Indigenous, and colonized people in struggles for decolonization and decoloniality over professional recognition, intellectual stardom, and academic success.

Rutgers University
New Brunswick, NJ, USA
Nelson Maldonado-Torres

Frantz Fanon Foundation
Paris, France

University of South Africa
Johannesburg, South Africa

University of KwaZulu Natal
Durban, South Africa

References

Césaire, A. (2017). *Journal of a homecoming/Cahier d'un retour au pays natal* (Gregson Davis, Trans.). Durham, NC: Duke University Press.

Dilanian, K. (2014). *2 psychologists helped run CIA interrogations*. Associated Press. Retrieved December 10, from https://apnews.com/article/96daff18a30249fcbd47b6c8c0ef414c

Fanon, F. (1967). *Toward the African revolution* (Haakon Chevalier, Trans.). New York: Grove Press.

Fanon, F. (2008). *Black skin, white masks* (Richard Philcox, Trans.). New York: Grove/Atlantic.

Fanon, F. (2018). *Alienation and freedom*. (Jean Khalfa, & R. J. C. Young, (Eds.), Steven Corcoran, Trans). London: Bloomsbury Academic.

Fish, J. M. (2014). Psychologists and torture. *Psychology Today*, Retrieved December 11, from https://www.psychologytoday.com/us/blog/looking–in-the-cultural-mirror/201412/psychologists-and-torture

Maldonado-Torres, N. (2008). *Against war: Views from the underside of modernity*. Durham, NC: Duke University Press.

Maldonado-Torres, N. (2017). Frantz Fanon and the decolonial turn in psychology: From modern/colonial methods to the decolonial attitude. *South African Journal of Psychology, 47*(4), 432–441.

Acknowledgments

We would like to thank Shahnaaz Suffla and Mohamed Seedat for the invitation to co-edit this volume, which is part of an exceptional three-book series, with the focus of our book being on decoloniality and epistemic justice in contemporary community psychology. We initially set out with the idea to recruit contributors from different geographical contexts, to capture the breadth and depth of this psycho-political project and to understand the divergent ways in which people have and continue to construct community psychologies around the world and in the context of significant global change. Despite this goal in mind, it was a challenging task for a host of reasons, yet we were still able to lay critical foundations for ongoing engagement with the decolonial turn and its expressions in psychological praxes situated in the knowledges, struggles, and realities of peoples navigating histories of slavery, colonialism, patriarchy, and other forms of oppression in the present. We understand these situated, culture-bound, ethically moored, historicized approaches as community-oriented psychologies within a broader set of paradigmatic shifts concerned with processes of liberation and rehumanizing. We are deeply appreciative of this opportunity for engagement and to be able to contribute to this critical task of epistemic refiguring.

We wish to thank the contributors for their patience as we navigated difficult times during the completion of the book. The COVID-19 pandemic had a significant impact on our working lives and capacity to finalize the volume sooner. Yet, we completed this task with the support of all of the contributors, despite the obstacles encountered by the massive changes to our everyday lives and having to navigate the online world and work from spaces in our homes and communities. We value the contributions and the critical, innovative, and politically informed examination of psychology and its complicity in coloniality, but also the possibilities for unthinking and delinking via epistemic disruption.

We are also deeply grateful to Nelson Maldonado-Torres for his incisive argument for decolonial psychology as counter-catastrophic science, in which he centers the writing of a foundational decolonial scholar—Frantz Fanon.

We also want to thank Roshani Jayawardana for supporting us through liaising with authors, reviewers, and formatting of manuscripts, all vital tasks to complete this book, as well as the members of the Community Identity Displacement Research group at Victoria University.

Finally, we would like to recognize those whose experiences and lives are primarily reflected and refracted within these chapters—those relegated to living on the periphery of society, the marginal, the excluded, the precariat, the *damné*—those whose lives reveal both the horrors of the darker underbelly of Western modernity each day, but who also simultaneously illuminate the spirit of human resilience, and the necessity for pursuing a project of critical and radical humanism that is attendant and resistant to the ongoing manifestations of coloniality and its variants in the lives of the majority world today.

Contents

About the Contributors

Rama P. Agung-Igusti is a PhD candidate at Victoria University, Australia, on the land of the Wurundjeri of the Kulin nation, where he completed a Bachelor of Psychological Studies (Honors), and is a member of the Community, Identity and Displacement Research Network. His research interests lie within the areas of migrant identity and belonging, racism, social movements, and community organizing.

Alison M. Baker is Associate Professor in Social Pedagogy at Victoria University in Melbourne, on the land of the Wurundjeri of the Kulin nation. Her research focuses on the implications of structures that produce inequality in the lives of various disenfranchised groups as well as those in positions of privilege. One strand of Alison's research focuses on young people's experiences of racialization and the implications for identity and belonging across contexts. She is interested in blending creative research methodologies and documentary techniques, particularly visual and sound modalities, to develop young people's sense of social justice and capacity for action.

Lesiba Baloyi is the Chief Clinical Psychologist, and Head of the Department of Clinical Psychology at Dr George Mukhari Academic Hospital and Sefako Makgatho Health Sciences University. He is the past president of the Forum of African Psychology and a member of the International Task Force of the Association of Black Psychologists (ABPsi), USA. His research interests include African epistemology and psychology, indigenous research methodologies, decolonizing and redesigning the psychology curriculum, postmodern thinking, Ubuntu philosophy, land and the impact of land dispossession on indigenous people, and indigenous knowledge systems in general.

Hugo Canham is an Associate Professor in the Department of Psychology at the University of the Witwatersrand in Johannesburg, South Africa. He is also the Assistant Dean of Graduate Studies in the Faculty of Humanities. His research explores waywardness and non-normative lives of various people who live on the margins of society. He works within the tradition of the black public humanities. His current book project is on the subject of black deathscapes. He teaches critical community psychology and qualitative research methodologies.

Urmitapa Dutta is an Associate Professor in the Department of Psychology at the University of Massachusetts Lowell. She is also affiliated with the Gender Studies, Peace and Conflict Studies, and Global Studies programs there. A feminist scholar activist, her program of research focuses on everyday violence, i.e., forms of direct, structural, and symbolic violence that become endemic to the social fabric. She uses critical qualitative methodologies to denaturalize oppressive conditions and to articulate experiences that are silenced by officially sanctioned narratives. Working alongside grassroots activists in Northeast India, her current research interrogates contested ideas and lived experiences of citizenship, migration, and belonging, while also exploring pathways for collective resistance and healing.

Michelle M. Fine is Distinguished Professor of Critical Psychology, Women's Studies, American Studies and Urban Education at the Graduate Center, CUNY. Fine taught at the University of Pennsylvania from 1981 to 1991, and then came to the Graduate Center. She has authored many "classics"—books and articles on high school push outs, adolescent sexuality—called the "missing discourse of desire," the national evaluation of the impact of college in prison, the struggles and strength of the children of incarcerated adults, and the wisdom of Muslim American youth. A pioneer in the field of youth Participatory Action Research, and a founding faculty member of the Public Science Project, Fine has been involved with a series of participatory studies with youth and elders, incarcerated and formerly incarcerated college students and youth working at the intersections of movements for educational, immigration, and juvenile justice.

S. Miranda Gierbolini is a Professor at the University of Puerto Rico. She received her PhD in Psychology at Temple University. She is part of the Graduate Social Community Psychology Programme at the University of Puerto Rico, Rio Piedras. She has received recognition for her research and publications on the topics of socio-economic grassroots community's development, mental health and disparities, violent culture, critical community psychology, and higher education, among others. She held the position of Chairperson of the Psychology Department, Faculty Representative on the University of Puerto Rico Board of Trustees, President of the Puerto Rican Professors Association, and the Puerto Rico Psychology Association. Among her recent publications are *The psy-complex and the legal economy* (translated) in Roman, M. (ed.) *Transiting the City Abandonment and Violence*, 2018, and with Ida de Jesus, *Thinking the Personal Debt Industry: Voices from Puerto Rico*, 2015. She was recently included in the publication of Martinez, Taboas, A., Roca de Torres, I., Boulon Jimenez, F., & Fernandez Ortiz, N. J. (2018). *Outstanding Psychologists: Professional Trajectory and Vision of Puerto Rican Psychology* (translated).

Shiloh Groot *(tribal affiliations include Ngāti Pikiao, Ngāti Uenukukōpako)* is a Senior Lecturer in Social Psychology at the University of Auckland. Shiloh is an interdisciplinary and Indigenous social scientist who works in the fields of Indigenous worldviews and communities, gender and sexuality, and homelessness and urban poverty. Shiloh is a long-standing member of the Tāngata Whenua Caucus for the New Zealand Coalition to End Homelessness (NZCEH), where they have been asked to advise on the expansion of research strategies that will inform the development of national policy and service provision.

Darrin Hodgetts is Professor of Societal Psychology at Massey University. Darrin identifies as a Pākeha New Zealander with familial ties to Ngai Tahu. Prior to his present post, Darrin held positions in community medicine at Memorial University in Canada, media psychology at the London School of Economics and Political Sciences, and community psychology at the University of Waikato. His recent books include (1) *Social Psychology and Everyday Life* (1st and 2nd eds), (2) *Urban Poverty and Health Inequalities*, (3) *Asia-Pacific Perspectives on Intercultural Psychology*, and the *Sage Handbook of Applied Social Psychology*. Darrin is an associate/review editor for the *Journal of Community and Applied Social Psychology* and *Palgrave Communication*, and an international adviser for *Health South Africa*.

Rolinda Karapu *(Ngāti Awa, Ngai Tūhoe)* is a Senior Manager, Waikato Women's Refuge Te Whakaruruhau. Rolinda has an extensive work history in the domestic violence sector and research with the Māori and Psychology Research Unit. Rolinda also works alongside central government Ministries to develop policies aimed at family violence reduction as part of her current role as Senior Specialist Practice Lead for the Integrated Safety Response Pilot in Waikato.

Pita King (*tribal affiliation—Te Rarawa*) is a Lecturer within the School of Psychology at Massey University, New Zealand. He is of Māori and Pākehā descent, hailing from the Northern New Zealand tribe of Te Rarawa. Pita's research interests include Indigenous and community psychology, homelessness and urban poverty, and the history and philosophy of psychology.

M. Brinton Lykes is Professor of Community-Cultural Psychology and Co-Director of the Center for Human Rights and International Justice at Boston College, USA. Her anti-racist, feminist activist scholarship focuses on: (1) rethreading life in the wake of racialized and gendered violence during armed conflict and in post-genocide transitions; and, (2) migration and post-deportation human rights violations and resistance. She has published extensively in refereed journals and edited volumes, co-edited four books, co-authored four others, and is co-editor-in-chief of the *International Journal of Transitional Justice.* Recipient of the Ignacio Martín-Baró Lifetime Peace Practitioner Award; the American Psychological Association's International Humanitarian Award; the Florence L. Denmark and Mary E. Reuder Award for Outstanding International Contributions to the Psychology of Women and Gender; and the Seymour B. Sarason Award for Community Research and Action, she is also a board member on several NGOs including Women's Rights International, Impunity Watch, and Grassroots International.

Nick Malherbe is a researcher at the Institute for Social and Health Sciences, University of South Africa and the South African Medical Research Council-University of South Africa Masculinity and Health Research Unit. His research interests include culture, politics, community psychology, and visual methods.

Bridgette Masters-Awatere *(tribal affiliations include Tūwharetoa ki Kawerau, Te Rarawa, Ngai Te Rangi)* is a Senior Lecturer in Kaupapa Māori psychology in the School of Psychology at the University of Waikato. She is also the Co-Director of the Māori and Psychology Research Unit and Convenor of the Community Psychology Graduate Programme—a professional practice program that leads to registration as a psychologist in New Zealand. Across Waikato University she provides academic leadership for the Mai ki Waikato (indigenous scholar network) and serves as a Māori research advisor on the University Human Research Ethics Committee (Health). Bridgette's research interests include evaluation research, Kaupapa Māori methodologies, indigenous well-being, and health inequalities in Aotearoa New Zealand.

Gloria G. McGillen is a PhD candidate in counseling psychology at the University of Missouri, USA. She is also a Graduate Clinician at the University of Missouri Counseling Center. She received her MA in Counseling Psychology from Boston College where she also served as a research assistant at the Center for Human Rights and International Justice. Her work addresses the lifelong sociopolitical development of adults and youth, with a focus on undocumented migrants, low-wage workers, and other vulnerable workers living in the U.S.

Cristina J. Montiel is a Professor of Peace/Political Psychology and has been teaching at the Ateneo de Manila University for more than 40 years. She received the 2010 Ralph White Lifetime Achievement Award from the American Psychological Association's Division of Peace Psychology. In 2016, she was recognized by the Psychological Association of the Philippines as their Outstanding Psychologist. Montiel is an international senior scholar in peace psychology. She was managing editor of the *Encyclopedia of Peace Psychology* (Wiley-Blackwell, 2012), and editor of the volume on *Peace Psychology in Asia* (Springer, 2009). She currently coordinates Ateneo de Manila's Political Psychology Research Lab. She has also been a consultant for the Philippine government's Commission on Human Rights and the Office of the Presidential Adviser on the Peace Process.

Linda Waimarie Nikora (tribal affiliations include Ngai Tūhoe, and Te Aitanga-a-Hauiti) was formerly the Director of the Māori and Psychology Research Unit in the School of Psychology at the University of Waikato in New Zealand. Since October 2017, she has taken up the position of Professor of Indigenous Studies at the

University of Auckland where she is also Co-Director of Ngā Pae o te Māramatanga, New Zealand's Māori Centre of Research Excellence.

Erwine S. Dela Paz holds a Master's Degree in Biology and is a doctoral candidate in the Psychology program of Ateneo de Manila. He is an active member of the Political Psychology Research Lab and specializes in text mining. He likewise teaches at Ateneo's Senior High School Department and currently heads its Research Department. He is also a part-time faculty at the Psychology Department of the Loyola Schools in Ateneo.

Kopano Ratele is the Director of the South African Medical Research Council Masculinity and Health Research Unit and Professor at the University of South Africa where he runs the Transdisciplinary African Psychologies Programme. He has published extensively and his books include *There Was This Goat: Investigating the Truth Commission Testimony of Notrose Nobomvu Konile* (2009, co-authored with Antjie Krog and Nosisi Mpolweni), *Liberating Masculinities* (2016), *Engaging Youth in Activism, Research and Pedagogical Praxis: Transnational and Intersectional Perspectives on Gender, Sex, and Race* (2018, co-edited with Jeff Hearn, Tammy Shefer, and Floretta Boonzaier), and *The World Looks Like This From Here: Thoughts on African Psychology* (2019).

Neville Robertson is a freelance community psychologist of Scottish descent. He recently retired from the University of Waikato where he taught papers in community psychology, evaluation research, qualitative research, and domestic violence. He now holds Associate status with the Māori and Psychology Research Unit at the University and continues to work as an evaluation researcher, particularly in relation to services responding to domestic violence and child abuse.

Mohi Rua (tribal affiliations include Ngai Tūhoe, Ngāti Awa, and Ngāti Whakaue) is the Co-Director of the Māori and Psychology Research Unit (University of Waikato), co-leader of the Mauri Ora (Human Flourishing) theme for Ngā Pae o Te Māramatanga (New Zealand's Māori Centre of Research Excellence), and Senior Lecturer in the School of Psychology, University of Waikato. Mohi's research focuses on Māori cultural concepts and practices, indigenous issues, health inequalities, poverty, and homelessness in Aotearoa/New Zealand.

Mohamed Seedat is Head of and Professor in the University of South Africa's Institute for Social and Health Sciences and was, until recently, the Director of the South African Medical Research Council-University of South Africa Violence, Injury and Peace Research Unit. His present research is focused on the psychologies underlying South Africa's ongoing and renewed struggles for a decolonized caring society, the social anatomy of public protests, and grassroots cultures of peace and safety. As part of a commitment to epistemic justice, Mohamed explores convergences in multiple counter-hegemonic enactments of psychology. He has supported community-engaged research, the capacitation of next generation socially engaged researchers and academic leaders, and the transformation of writing cultures in the academy. His body of work contributes to cross-disciplinary compassionate emancipatory scholarship for the twenty-first century. Mohamed also writes about the meanings and enactments of emancipatory community practices and the messiness inherent to resisting exclusionary knowledge practices.

Puleng Segalo is an Associate Professor at the University of South Africa. She is based in the Department of Psychology where she teaches both Social and Community Psychology. She is the current president of the Forum of African Psychology and the coordinator of the UNISA Decoloniality project. Her research work and publications cover a wide range of areas including gendered experiences of women in various aspects of life, Critical Participatory Research Practices and Knowledge Production, Power, and Decoloniality.

Christopher C. Sonn is Professor at Victoria University, Melbourne, Australia, on the land of the Wurundjeri of the Kulin nation. He is a fellow of the Institute of Health and Sport and teaches in the Applied Psychology Programme in the College of Health and Biomedicine. His research is concerned with understanding and changing dynamics of oppression and resistance, examining structural violence such as racism, and its effects on social identities, intergroup relations, and belonging. He holds a Visiting Professorship at the University of the Witwatersrand, South Africa. He is co-editor of *Creating Inclusive Knowledges* and co-author of *Social Psychology and Everyday Life,* and Associate Editor of the *American Journal of Community Psychology* and *Community Psychology in Global Perspective.*

Garth Stevens is a Professor and Clinical Psychologist at the University of the Witwatersrand in South Africa. His research interests include foci on race, racism, and related social asymmetries; critical violence studies; and historical/collective trauma and memory. He has published widely in these areas, both nationally and internationally, including co-editorships of *A 'race' against time: Psychology and challenges to deracialisation in South Africa* (UNISA Press, 2006) and *Race, memory and the apartheid archive: Towards a transformative psychosocial praxis* (Palgrave Macmillan, 2013). At present, he is a member of the Academy of Science of South Africa (ASSAf), serves as the Dean in the Faculty of Humanities at the University of the Witwatersrand, and is President of the Psychological Society of South Africa (PsySSA).

Shahnaaz Suffla is an Associate Professor at the University of South Africa. Her research interests draw from the intersections of critical African, community, and peace psychologies and are located within liberatory philosophies and epistemologies. Her thinking and scholarship is influenced by the vision of research as a transforming, humanizing, and decolonizing enterprise. Her research interests include a focus on community-engaged interventions in contexts of structural and epistemic violence, participatory engagement as a site of activism, resistance and social change, and Africa-centered knowledges and knowledge-making.

Mary Watkins is Chair of the Depth Psychology doctoral program at Pacifica Graduate Institute (Carpinteria, CA) and Co-Chair and Co-Founder of its Community, Liberation, Indigenous, and Eco-Psychologies specialization. She is the author of *Mutual Accompaniment and the Creation of the Commons, Waking Dreams, Invisible Guests: The Development of Imaginal Dialogues;* and the co-author of *Toward Psychologies of Liberation, Up Against the Wall: Re-Imagining the U.S.-Mexico Border,* and *Talking with Young Children About Adoption.* She is currently engaged in psychosocial accompaniment of asylum seekers in the U.S. and in liberatory education in prison contexts.

Tracking the Decolonial Turn in Contemporary Community Psychology: Expanding Socially Just Knowledge Archives, Ways of Being and Modes of Praxis

Christopher C. Sonn and Garth Stevens

We live in perilous times. Some of the defining features of the current historical juncture is that it reveals sustained forms of inequality, is increasingly characterised by new modes of alterity, and encourages levels of impunity and imperviousness to organic pressures for social change and egalitarianism today. But, as a contrapuntal response to this, we witness a counter-ethics directed towards minimising social hierarchies, enactments in the service of the public good, a drive for historical analyses of the failures of socio-political and economic systems, and new modes of praxis to remediate these systemic forms of dereliction.

Central to such a response is how we recalibrate knowledge about strategic relations of power, social systems, and subjects. To this end, the current volume, as part of a series[1] on decoloniality and community psychology, set out to critically interrogate the biases in Western modernist thought in relation to community and related applied psychologies, and to illuminate and consolidate current ontological and epistemic alternatives that contribute to the possibilities of emancipatory futures in various local and global contexts. While we set out with a focus on what is commonly referred to as the Global South, which typically refers to regions outside of Europe and North America, it became clear that decoloniality and decolonial praxis has no geographical limits. Mignolo and Walsh (2018) note that decolonial thinking and praxis:

[1] This series includes 'Decolonial Feminist Community Psychology' (Boonzaaier & van Niekerk, 2019) and 'Decolonial Enactments in Community Psychology' (Kessi, Suffla, & Seedat, 2021).

C. C. Sonn (✉)
Institute of Health and Sport, Victoria University, Melbourne, VIC, Australia

University of the Witwatersrand, Johannesburg, South Africa
e-mail: Christopher.Sonn@vu.edu.au

G. Stevens
Department of Psychology, University of Witwatersrand, Johannesburg, South Africa

G. Stevens, C. C. Sonn (eds.), *Decoloniality and Epistemic Justice in Contemporary Community Psychology*, Community Psychology,
https://doi.org/10.1007/978-3-030-72220-3_1

> …is present in the civilizations of the 'Souths', in Asia, the Pacific, the Arab world, and Africa as well as Latin America; the South that Kumar poetically describes as insurrection of subjugated knowledges, history, memory, and new political imaginaries. Decolonial praxis traverses the Souths, including the Souths in the North (p. 45).

Decoloniality as Paradigmatic

The decolonial turn, with a focus on epistemic interventions, is frequently cited as only emerging in the literature in the late 1990s and early 2000s with key writings by Walter Mignolo, Ramon Grosfoguel, Sylvia Wynter, Enrique Dussel, Nelson Maldonado-Torres, and Linda Tuhiwai Smith, amongst others. However, the landscape of decoloniality as a site of intellectual engagement has existed for a much longer period, with impressive lineages and genealogies amongst a range of intellectuals such as Aimé Césaire, Albert Memmi, Frantz Fanon, Paulo Freire, Steve Biko, Amilcar Cabral, Kwame Nkrumah, Patrice Lumumba, Robert Sobukwe, and still others would argue, even in the earlier writings of thinkers such as W.E.B. Du Bois (Stevens, 2016). Importantly, as Quijano (2000) and other scholars (e.g., Lugones, 2010) have noted, decolonizing knowledge is not new and has roots as long as colonization. Maldonado-Torres (2011) notes that the decolonial turn:

> …has long existed in different ways, opposing what could be called the colonizing turn in Western thought—the paradigm of discovery and newness that also included the gradual propagation of capitalism, racism, the modern/gender system, and the naturalization of the death ethics of war (Maldonado-Torres, 2011, p. 1).

The decolonial turn is thus paradigmatic—it is not comprised of a singular theoretical school or orientation. As noted by Maldonado-Torres (2011), the decolonial turn, "… points to a family of diverse positions that share a view of coloniality as a fundamental problem in the modern (as well as post-modern and information) age, and of decolonization or decoloniality as a necessary task that remains unfinished" (p. 2). He continues, stating that for those who seek to advance the discourse of decolonization beyond the dialectics of identity and liberation, recognition and distribution, we have to add "the imperative of epistemic decolonization, and in fact, of consistent decolonization of human reality. For that one must build new concepts and be willing to revise critically all received theories and ideas" (p. 4).

Given this conception, it is therefore not surprising that decolonial theory is thriving and has significant purchase power in analyses of colonial and post-colonial contexts, because its overarching premise rests in the critique of Western modernity or Western Enlightenment, its terrors and its intractable internal contradictions (Maldonado-Torres, 2016). Whilst certainly overlapping with other lenses in psychology today, including Critical Psychology, Black Psychology, Feminist Psychology, and Liberation Psychology, for example, decolonial theory is much more broadly and paradigmatically directed towards understanding the historico-ontological effects of coloniality as a system, on contemporary social arrangements and subjects today. For decolonial theorists, the entire Western modernist project and its associated colonial matrix have shaped the very nature of power, knowledge, and being. Quijano (2000) coined the term "coloniality of power" to name this matrix, "the continuities in the so-called 'post-colonial era' of the social hierarchical relationships of exploitation and domination between Europeans and non-Europeans built during centuries of European colonial expansion emphasizing cultural and social power relations" (p. 9). For Quijano (2000), amongst others, decoloniality is fundamentally poised to engage systems of hierarchy, systems of knowledge, and systems of culture associated with Western modernity.

As noted previously, decolonial theory and praxis is heterogeneous—it is not a monolithic body of knowledge but has multiple offshoots and genealogies. In North America, it gained traction through Garveyism and movements such as the Harlem Renaissance, and in response to a history of chattel slavery, the Jim Crow Laws, and movements stressing a return to Africa; in the Caribbean the history of the middle passage and its shaping of the entire social and political terrain, together with important work on Negritude helped to propel these ideas; in Latin America, successive dictatorships prompted a turn to critical and radical pedagogies of freedom, dependency theory, liberation theology, liberation philosophy and subsequently liberation psychology; in Aotearoa and Australia, the focus was on indigeneity and the defence of Aboriginal and Torres Strait Islander, and Maori peoples, with a focus on indigenous ways of knowing, doing and being; in Africa the independence movements of the mid-twentieth century heightened calls for Afrocentricity, African Socialism, and Pan Africanism; and in India, the Independence Movement and the Quit India Movement helped to craft a further iteration of decolonisation and decoloniality. Historically, these developments have all profoundly influenced psychology as a discipline and practice, and it is these more marginal epistemic trajectories that are evidenced in this volume, and that are re-centred as part of the project of onto-epistemic justice that is at the heart of decoloniality (Stevens, 2018a).

The Centrality of Epistemic Justice in Decoloniality

The pursuit of epistemic justice is evident across many fields, but it is central to addressing the coloniality of knowledge, and relatedly, the coloniality of power and being. Visvanathan (2009) uses the term cognitive justice and defines it as "the right of different forms of knowledge to co-exist but adds that this plurality needs to go beyond tolerance or liberalism to an active recognition of the need for diversity" (para. 21, line 2–3). Cognitive justice in some ways can be viewed as a counter project to what Boaventura de Sousa Santos (2016) has described as the "sociology of absences", which refers to silences around experiences, how silences are created, and practices that make different knowledges and struggles invisible within the global knowledge system. Absences are produced through different logics, including through the process of legitimizing the canon and producing a monoculture of knowledge and not recognizing anything outside of this as existing. Another logic is based on the "monoculture of the naturalization of differences" (p. 274), of which racial and sexual classification are salient examples of this thinking. This logic produces relations of domination and the non-existence of those deemed naturally inferior. In other fields, scholars have advocated for the examination of ignorance, to understand the role of power and privilege in the production and maintenance of gaps in knowledge practices (see Sullivan & Tuana, 2007).

The concept of epistemic injustice has also been theorized by Fricker (2007), who refers to the type of harm done to someone in their capacity as a knower. Fricker (2007) describe two types of epistemic injustice, namely testimonial, and hermeneutical injustice. Testimonial injustice is the prejudice that results in a speaker receiving *more* credibility than they otherwise would have, or it results in them receiving *less* credibility than they otherwise would have. An example of this is how the lived experiences and knowledges of some groups are not valued, or are perceived as not manifesting a strong evidence base for credibility. Hermeneutical injustice occurs when there is a deficit or absence in our shared tools of social interpretation (the collective hermeneutical resources) which constrains social and psychic space simultaneously. For example, marginalized and oppressed social groups are at a disadvantage in making sense of their distinctive and important experiences, given a deficit in the collective hermeneutical resources for making sense of such experience. From these perspectives, the pursuit of epistemic justice is about expanding our ecology of knowledge, to decentre, unlink, retrieve, and recentre ways of knowing, doing, and being that have been violently devalued and omitted through the coloniality of power, and that continue to be contested 'at the borders' of the colonial matrix that characterizes contemporary planetary entanglements (Ndlovu-Gatsheni, 2020).

A Moment of Social Reckoning and Transformation?

The present historical juncture has seen the forced changes ushered in around the globe by the COVID-19 pandemic and its disproportionate sequelae amongst the *damné* of the world. The struggles for Indigenous Sovereignty, the Black Lives Matter Movement, and climate justices represent a coalescing of 'crises' and an urgent call to embark on a project of epistemic reconstruction, given the malaise of our time, as well as the possibilities for emancipatory futures that are simultaneously revealed. In his *Prison Notebooks* in the 1930s, Gramsci (1971) referenced both "morbidities" and the "interregnum" as key ideas to define such crisis moments in which a social reckoning and remediation of sorts were possible. Of course, Gramsci was appropriating the idea of the "interregnum" that was normally utilized to define the period of political transfer of power from one monarch to another, but utilized it to describe:

> the extraordinary situations in which the extant legal frame of social order loses its grip and can hold no longer, whereas a new frame, made to the measure of newly emerged conditions responsible for making the old frame useless, is still at the designing stage, has not yet been fully assembled, or is not strong enough to be put in its place (Bauman, 2012, p. 49).

This in-between moment, where there is also the confluence of various social crises, is a period where one arrangement of hegemony is waning, but before the emergence of another. This is akin to how Arundhati Roy (2020, n. p. number) described the pandemic as "… a portal, a gateway between one world and the next" and that "have forced humans to break with the past and imagine the world anew".

Stevens (2018b) notes that morbid symptoms characterize much of our psychological, social, and political life today. Citing writers such as Humphrys (2015) and Mair (2013), he argues that there is a profound historical shift taking place, characterized by the decline in the mass-based politics that persisted for most of the twentieth century. Here, there is detachment from, and hostility to, politicians, parties, and the political process, expressing itself in short-lived bursts of protest, electoral volatility, and political crises. But there are also political projects from across the ideological spectrum that trade on an appeal to this mood for their political ends—mobilising widespread disenchantment with politicians and the political process into new formations promising a fundamentally new politics. Stevens (2018b) notes that rather than analysing these moments as a mere pathologisation of the citizenry as racist, conservative, populist, etc., we should not ignore the dissolution of the social base of longstanding political arrangements—an entropy in the political order premised on the ever-increasing detachment of classes from their political representatives. Today, we witness such reverberations across the globe—a decline in the confidence in ruling parties and modes of governance, an upsurge in social movements and protests, broad-based civil society movements challenging the established political order, and of course, student protests at many institutions of higher learning.

More than ever, there is perhaps a range of decolonial impulses across the globe—in Africa, various formations have mobilized around issues of landlessness, service delivery, and around the transformation of higher education; in Latin America, we have seen coalitions between the precariat and working-class formations in insurgent citizen responses to the increasing centrism of ruling parties; Portugal, Spain and Greece have been the homes of mass protests and a groundswell against the austerity measures of the Eurozone; and the USA has seen its fair share of the occupy movements, #BlackLivesMatter, and remains embroiled domestically and abroad in debates on the virtues or failures of its aggressive foreign policies in various parts of the world (Stevens, 2018b). Of course, these moments of resistance are also accompanied by illustrative exemplars of failed or weakened states, the presence of increasingly fascist groups and dictatorial regimes, the continuation of state-sanctioned hostilities, and so forth. It appears that Francis Fukuyama (2006) was perhaps somewhat premature when he declared *The End of History* and the victory of Western liberal democracy, given the resistances to it, the decline of false consensus and the breach of ignorance contracts, that are all central ligaments that connect systems of inequality and violence amongst the citizenry of the world—all representing a fundamental crisis of Western modernity and its associated forms of knowledge production, power and ways of being (Stevens, 2016). Furthermore, as noted by Ndlovu-Gatsheni (2020), the COVID-19 pandemic has accentuated multiple crises, including in the domains of the ecology, politics, and health, as well as a crisis in knowledge—an epistemic crisis. He notes that the current challenges have produced a crisis in our knowledge systems, in which we are poised but not yet ready for thinking differently, and in ways that potentially centre the planetary entanglement of the human and beyond human. It is in this potential tipping point context, that psychology itself has once more begun to interrogate its own ontological and epistemological complicities with coloniality; to assess its emancipatory possibilities and limitations; and to reimagine itself in relation to other modes of praxis, even if this involves the abandonment of certain disciplinary sacred cows.

Psychology's Decolonial Turn

Community psychology, and more recent iterations of critical community psychology, because of their underlying value base, including cultural relativity and diversity, person-environment co-constitution, social justice, and power, as well as their potential for criticality (Coimbra et al., 2012; Kagan, Burton, Duckett, Lawthom, & Siddiquee, 2019; Rappaport, 1977), have perhaps always been positioned to embrace processes of epistemic reconstruction and justice in the service of emancipatory futures. If there is a single claim that can be made about the world today, then it is that it is fundamentally based on unevenness, inequality, and marginalization (Keeley, 2015; Pickett & Wilkinson, 2015; Therborn, 2009). If community, critical, and related expressions of the discipline are to realise the fundamental goals and commitments to contribute to social justice and transformation they will have to speak to certain key questions—and indeed in many instances they do—about ontology, ethics, methods, theory, content, knowledge production and dissemination processes, and the re-centering of epistemic subjects/agents that are generally marginalized (see Seedat et al., 2017). To this end, they have and must continue to draw upon and expand on the range of critical psychologies that currently exist, ensuring that such a decolonial project in psychology transcends the politics of identity and recognition with the attendant re-distributional regimes that normally accompany them.

There are several examples in community and related applied psychologies that are engaging with the decolonial turn (e.g., Boonzaier & van Niekerk, 2019). One recent example is Sunil Bhatia's (2018) book, *Decolonising Psychology: Globalization, Social Justice, and Indian Youth Identities*, in which he examined how young people navigate and negotiate economic, social, and cultural changes brought about by globalisation in their everyday lives. Drawing on post-colonial theory and narrative psychology, he highlights how the youth construct selves and make sense of tensions between home culture, values and habits, and Westernised culture. Bhatia raises questions about the exporting of Euro-American scientific psychology into India, the absences about ways of being, culture, and traditions that are produced through this process, and the resulting universalizing and colonizing effects of this psychology. The engagement with decolonization has entailed renewed critiques of 'ruling' psychology rooted in universalist assumptions of a psychic unity of humankind within a hegemonic Western Eurocentric framework. In this framework, human civilization originates in Western Europe, including the exclusionary and hierarchical conceptions of what it means to be human. These ideological conceptions of the human center European 'man', what Sylvia Wynter referred to as Man 2, the modern-day bio-economic/liberal humanist entity. This conception of the human has been central to colonialism and the coloniality of power and have endured in the construction and production of psychological research and practice (see Teo, 2005).

Many scholars have critiqued various branches of applied psychology for its complicity and commitment to 'epistemologies from nowhere', that is, for ahistorical, acultural, and decontextualized conceptions of people that have contributed to the construction of people relegated to the zone of non-being (e.g., Adams, Dobles, Gómez, Kurtiş, & Molina, 2015; Kessi, 2017; Montero, 2007; Ratele et al., 2018, Ratele, 2019; Sonn, Rua, & Quayle, 2019). These challenges go way beyond the phenotypical features of the community and applied research, calling for a much deeper unsettling of the genotype of our disciplines and to imagine psychologies otherwise and pluriversal. These are by no means new or easy challenges but concerns the centering of knowledge, ways of being, knowing, and doing that have been ignored, silenced, and excluded within the longer histories of empire and colonization. Decoloniality is not the search for universals, it is indeed the unlinking and rethinking, the shifting away from universality and interlocking systems of domination that continue to serve some communities over others and that have produced the crises that we currently face. Mignolo and Walsh (2018) write that:

> Decoloniality is understanding how coloniality affects our thoughts and feelings about ourselves, how we make sense of reality, how we view our relationships with other humans, other species, or the natural environment, and the constraints or possibilities we perceive there to be all around us (p. 140).

This definition is deeply relevant to community and applied psychology and our disciplines must play a role in critically examining our complicities with coloniality.

There is a growing number of journal publications dedicated to the topic of decolonization and decoloniality in community and applied social psychology, including a special section of the *Journal of Social & Political Psychology* (volume 3, issue 1, 2015), the *South African Journal of Psychology* (volume 47, issue 4, 2017), and the *American Journal of Community Psychology* (volume 62, issue 3–4, 2018). These special issues have brought into focus the possibility of decolonial community psychologies toward a decolonial standpoint (Reyes Cruz & Sonn, 2015), a central project for expanding our ecology of knowledge, pluriversality, towards epistemic justice by interrogating and disrupting colonial legacies that continue to be entrenched in ways of knowing, doing and being in the world. More specifically, Carolissen and Duckett's (2018) special issue on decolonial pedagogies in community psychology distills key markers of decolonial pedagogy from a collection of 15 papers. They list several markers central to decolonial pedagogy, some of which are listed here:

- Disrupting the privileging of Euro-American/Western epistemologies;
- Reframing pathologized accounts of marginalized peoples;
- Reclaiming and reframing the erasure of histories;
- Deconstructing colonial discourse, and inserting counter-narratives;
- Centering indigeneity and indigenous knowledges in the curriculum;
- Foregrounding the politics of knowledge production;
- Pedagogical tools employed as catalysts for reflexivity; and
- Drawing on ecologies of knowledge, appropriate to the context.

These markers show how people are concretizing praxes toward the decolonial turn and imaging decolonial community psychology 'otherwise' (Walsh, 2015). For Walsh (2015) otherwise means ways of: "being, thinking, knowing, sensing, feeling, doing and living in relation that challenge the hegemony and universality of capitalism, Euro-centred modernity and the Western civilizatory logic, including its anthropocentrism and binary-based foundation" (p. 12). Fernández, Sonn, and Carolissen (2019; Fernández, Sonn, Carolissen, & Stevens, 2021) distilled insights from workshop discussions about how people in different countries are engaging with and making meaning of the decolonial turn in pursuit of decolonial community psychology otherwise. They noted the centrality of positionality and location, ethics, and a turn away from orthodoxy in people's enactment of decolonial praxis. These ideas bring into focus both the politics of location and representation. Specifically, Fernandez et al. (2019) identified orientations and principles that centre local and indigenous knowledge creation in partnership, socio-historical ethical consciousness, mutual accountability, and unsettling subjectivities of power/privilege.

The current volume is part of the process of expanding our ecologies and archives of knowledge, and embracing pluriversality, to add to the growing stock of resources aimed at advancing epistemic justice. To this end, the current book, along with the others in this series, offers exemplars in community psychology praxis across the globe that: speak to standpoint methodologies in which the experiences of the majority of the global population are more accurately reflected; recover forms of subordinated thought or intellectual traditions; address key social issues such as the on-going racialization of the globe, gender, class, poverty, xenophobia, sexuality, violence, diasporas, migrancy, etc.; embrace forms of knowledge production that involve the co-construction of new knowledges across the traditional binary of knowledge producers and consumers and take seriously matters of curriculum transformation; and finally, contribute to re-imagining epistemic subjects that have frequently been marginalised so that we craft alternative ways of being in the world that have greater empowering and emancipatory potential. Fundamentally, such decolonial approaches are about the relationship between various sciences and society, as modes of cultural production, that have the prospects for contributing to a pluriversalism and new forms of critical humanism in a sustained manner.

Centre, Periphery and the Elasticity of Coloniality

Before we distil insights from the contributions it is important to highlight the efforts involved in creating such a volume and the major obstacles that are part and parcel of the inequality that characterises the knowledge economy and the differential access to resources, variations in knowledge production practices, foci of research and outputs, and the matter of language and translation. While we reached out to a wide range of people across the global South and South in the North, only a relatively limited number of people were able to complete full papers that have been included in this volume, representing contributions from authors in several different countries including Australia, New Zealand, Puerto Rico, South Africa, the Philippines, and the United States. So, we acknowledge challenges and obstacles to participation and the partiality of the contributions offered in this volume. Yet, while partial, the chapters make meaningful contributions, showing how various scholars, activists, and practitioners in distinct locales and contexts have enacted visions and created practices toward decolonial horizons for applied and community psychologies. The chapters however also raise important questions for us to consider about the decolonial turn, decoloniality, and risks associated with simply replacing one discourse with another. It is vitally important to ask questions such as: Is decoloniality the new way of engaging with critical work and is it at risk of becoming a free-floating signifier within a defence of psychology that comes to imply criticality, but with such an amorphous set of meanings that it is rendered meaningless and impotent as a paradigmatic resource? Is decoloniality as a concept being appropriated by various centers away from the margins? How are the margins shifting and being redrawn concerning the decolonial turn? Is there a new politics of intellectual and experiential extractivism and appropriation that we need to be mindful of? These are significant questions that are ultimately about enduring relationships between the centre and the periphery, the global knowledge economy, the university, and the risks associated with circuits and flows of knowledge within uneven and unequal systems of knowledge production and dissemination that have significant elasticity and recalcitrance. Thus, as we embrace and engage with the decolonial turn, we must ask the critical questions about epistemic location and knowledge production, from where and to where, and ultimately how knowledge is produced, legitimized, circulated, and returned to its place of enunciation and by whom?

Praxis Co-ordinates for Decoloniality, Knowledge Production and Epistemic Justice

The various chapters in the volume show ways in which people in different contexts are engaging decolonisation in the pursuit of epistemic justice and liberation. Linda Tuhiwai Smith (1999) wrote:

> Decolonisation is a process which engages with imperialism and colonialism at multiple levels. For researchers, one of those levels is concerned with having a more critical understanding of the underlying assumptions, motivations, and values which inform research practices (p. 21).

Nelson Maldonado Torres (2016) several years later referred to decolonisation as:

> efforts at rehumanizing the world, to breaking hierarchies of difference that dehumanize subjects and communities and that destroy nature, and to the production of counter-discourses, counter-knowledges, counter-creative acts, and counter-practices that seek to dismantle coloniality and to open up multiple other forms of being in the world (p. 10).

In these chapters we have exemplars of how various actors are, in their respective social-cultural locations, producing perspectives, processes, actions, and ways of doing that turn 'away from Western modernity/coloniality'. In community psychology, the turn *away from* often involves interrogating long-standing worldviews, taken for granted notions and practices that have been solidified in Euro-American centric psychologies and that have been and continue to be complicit in coloniality. In his turn towards the South, Tod Sloan (2018) eloquently suggested that: "Decolonial theorists help us see how Reason rationalized genocide, slavery, and exploitation; how Coloniality still seriously constrains human possibilities long after geopolitical decolonization; and what forms of praxis will be necessary if we are to achieve just societies" (no page number). In this volume, we see ways of knowing, doing, and being that resonates with the local constructions and enactments of decoloniality in forging community psychologies that are attentive to our global entanglement and continued re-articulation of matrices of power, which are at the centre of the various crises facing the planet. Below we offer a tentative formulation of co-ordinates for decoloniality and epistemic justice evident from the chapters in this volume. Decoloniality as expressed in these chapters means interrelated practices, processes, and knowledge production activities towards transformative social change. We have organised some of the key ideas into three interrelated categories that refer to practices and principles that are aimed at enacting decoloniality and epistemic justice in community and applied psychologies: (1) Archival reclamation and expansion for onto-epistemic disruption, (2) Knowledge from below and centring voices and subjects from the margins as epistemic justice, and (3) Modes of praxis that encompass critical mutual accompaniment, acknowledge intersectional identities, and are premised upon dialogical ethics.

Archival Reclamation and Expansion for Onto-Epistemic Disruption

Archival reclamation, expansion and interrogation are central to the decolonial project (see Bulhan, 1985; Stevens, 2017; Stevens, Duncan, & Sonn, 2010). These practices signal how various authors have and are turning away from Western modernity/coloniality. These practices of reclamation, expansion and interrogation signal disruption to the "sociology of absences" (de Sousa Santos, 2007, 2016) and epistemological ignorance in the pursuit of epistemic justice. In this volume, Seedat and Suffla argue that it is imperative to engage with the multiple knowledge archives that have been neglected or made appendage to the Western-centric ontological and epistemological frames that continue to shape the underlying assumptions of community psychology in many countries. They draw on the Black Consciousness philosophy advanced by Steven Bantu Biko to decentre the ontological and epistemological underpinnings of Western-centered psychology and offer at the same time a decolonial point of departure based on collectivist and humanizing ontology for the enactment of emancipatory projects. Suffla and Seedat note that:

> The Black Consciousness archive offers us profound and authoritative insights on the ever-urgent imperative of transformation in (community) psychology. Biko's culturally constitutive ontology, an element of which we conceptualise here as collectivist and humanising ontology, seeks to break from the colonial conception of being, and challenge its associated erasures and negations of black personhood (p. 35).

Ratele and colleagues also ask pertinent questions about the possibilities for decolonial Africa(n)-centred community psychology. Through an innovative conversational approach, they name the turn away from coloniality of power. They state that: "Coloniality inheres in Euro-American centralizing racialized economic world power but, equally important, it is the quintessence of the very categories of thought, in work, in social relations, and in identity-making" (p. 44). They explore challenging questions about how epistemological oppression continues and identify the mechanisms such as sub-oppression through which teachers, researchers, and others become complicit in the perpetuation of coloniality. Ratele and colleagues examine pertinent questions about how to imagine decolonial Africa(n) community psychology, and pathways toward "imaginative horizons by which a world-centered community psychology that is within, for, by and of Africa is able to embrace decoloniality" (p. 55). Along the same lines of epistemic disruption, Canham and his colleagues challenge the Western canon, which often reinforces absences and epistemological ignorance, and draw on their experiences and reclamation of African epistemologies to enact epistemic justice through the pedagogical encounter. They go on to reflect on their personal challenges, concerns, discomforts and attempts at re-imagining and engaging with decolonial pedagogy, and convey their views about the ethical responsibility they carry as African scholars pursuing epistemic justice.

Fine also calls up the buried legacy of twentieth-century community research contributions. She highlights how the significant, often neglected, knowledge created by scholars like W.E.B. Du Bois is counter to the colonising archive produced through racist, sexist, and zero-point epistemologies. Fine resurrects "activist-scholar ancestors" who have suffered "death by canon" and argue for an approach to research and community-engaged action that recognizes the circuits through which power, privilege, and dispossession circulates with disastrous consequences for those at the sharp end of the wedge. Fine and others in the volume urges us to push beyond binary conceptions of self and other, past and present, theory and practice, towards an engagement with the complexity of the borderlands inspired by the seminal work of Chicana Feminist theorist activist-scholar, Gloria Anzaldúa (1999) and also the decolonial feminist work advanced by Lugones (2003, 2010; see also Boonzaier & van Niekerk, 2019). Contact zones are constructed in borderlands, where differently positioned subjects come together to express desires, connect, deconstruct and de-ideologise dominant discourses, and form and forge new alliances and commitments in efforts to promote justice. "Our field—critical community psychology—is well positioned to catalogue, with humility, the lives, communities and beliefs being slaughtered; the wilful subjectivities, solidarities and forms of resistance being borne" (Fine, 2015, p. 75).

More contemporary retrieval and centering are both captured in the chapter by Lykes and Gillen, who trace the work of Ignacio Martín Baró, a key proponent of liberation psychology. Martin Baró's contributions sit within a broader paradigm of liberation that has links with Dependency Theory, Liberation Philosophy, Critical Pedagogy, and Sociology that emerge across Latin and Central American countries (Montero, Sonn, & Burton, 2017). Liberation psychology is the branch of this paradigm in the discipline of psychology and was initially articulated by Martin Baró and developed by different proponents including Maritza Montero (Montero, 2007; Montero & Sonn, 2009) and others in the English-speaking world (Burton & Kagan, 2005). The authors outline the significance of liberation psychology and Martin Baró's drive towards a new horizon, epistemology, and praxis, which resonate powerfully with the decolonial turn. Martin Baró's work has presented a significant challenge to psychological orthodoxy, yet even though many of his ideas have been taken up by scholars, practitioners, and activists, Lykes and Gillen show that these ideas remain at the margins of the field, but importantly, how these ideas also endure and are vital to rehumanizing psychological praxis in the pursuit of emancipation.

Knowledge from Below, Centring Voices/Subjects from the Margins, and Epistemic Justice

Archival reclamation and reparation are central to the onto-epistemic disruption of Western modernity and ushers in new ways of working that includes methods, practices, pedagogies, and praxis rooted in relational ethics, and the embrace of plurality as vital to the expansion of the ecology of knowledge. Several chapters in this volume illustrate an engagement with practical and theoretical challenges in praxis in the here-and-now that is cognizant of history, structure and experience. Dutta mobilises the everyday as an analytical category in her ethnographic work that centres on the experiences of communities in Garo Hills in India, confronting the challenges of ethnic violence. These lived experiences are counter-stories that rupture dominant discourses and grand narratives which often naturalize 'ethnic violence'. However, by centring the lived experiences of young people in Garo Hills within a broader history, structure and circuits of power, the everyday struggles open up pathways to locally-situated constructions of peace-building that denaturalize hegemonic victim-blaming conceptions of ethnic violence. Montiel and De La Paz also discuss the complexities of peace-building initiatives in the post-colonial Philippines with a focus on how colonial practices of pacification, a form of coloniality of power, is expressed in the present and how even peace and social justice initiatives can be uneven. The authors note:

> The word *peace* may have acquired a subjugating connotation, associated with historical memories of colonial pacification drives, and post-colonial efforts to mute local resistance efforts against external domination. Our study shows how dual concepts of a post-colonial pacification narrative by a central government, and a decolonizing storyline by peripheral or ethnic groups, are expressed during intra-state peace negotiations (p. 224).

In a different colonial context, Gierbolini traces the history of diverse forms of grassroots community work that are participatory, collective and localised in Puerto Rico. She outlines the development of critical community psychology that has been developing over time and that is anchored in a critical and liberation perspective or framework. Decoloniality for their group:

> suggests action to address the colonial geopolitical context, the knowledge, and the subjectivity triad. All three concepts constitute decoloniality. We have tried to understand the situation in Puerto Rico contributing to create alternative subjectivations—the process of creating alternative subjectivities in the social construction of a decolonial being. A liberated being (p. 129).

Participatory methods and a relational action-oriented approach to action-theory-action cycles are located within a participatory action paradigm, which for Gierbolini is now deepening as the group embraces ecologism in the face of the planetary crises that are expressed in natural disasters that impact the island nation and the lives of people in Puerto Rico. Ortiz Torres (2020) provides a further example from Puerto Rico of participatory community psychological practices with and alongside communities, which support self-determination and autonomous organizing aimed at individual and collective wellbeing and demand for rights. In a different part of the world, Rua et al. use Tuck and Yang's statement that "decolonisation is not a metaphor" and document their work, which spans several decades, to carve out a community-based participative approach to decolonising research. This approach is one that is culturally situated and evolving from Kaupapa Maori perspectives within Aotearoa/New Zealand.

This Indigenising cultural approach to decoloniality rooted in Kaupapa Maori is akin to calls from First Nations people to indigenize psychology, which includes a cultural worldview and ways of being derived from the cosmologies and knowledge that have sustained communities for centuries (see Bhatia, 2018; Ciofalo, 2019; Dudgeon & Walker, 2015). Rua and colleagues highlight several cultural values and principles rooted in Maori culture that are central to their iterative, generative, place and culture-bound praxis, sensitive and responsive to the needs of communities. As they note: "Working collaboratively with such activists to realise our obligations towards intergenerational efforts to achieve the survival and thriving of Māori today and into the future is of primary concern. It is also a fulfilling way to spend one's time" (p. 190). In Australia, Rigney (1997) articulated principles of Indigenist research to include that it privileges indigenous voices, has political integrity, is aimed at resistance, and is central to emancipation. Martin and Mirraboopa (2003) also outlined principles that are anchored in Aboriginal peoples' world views, knowledge and realities, and the need to privilege Aboriginal peoples' realities, honouring social mores and practices on different lands, and the understanding of context in shaping experiences. Central to this writing was the articulation of different ontologies, cosmologies, and ways of being rooted in relationality and connected between people and the natural world.

Endogenous participatory and socio-cultural approaches have deep roots in majority world theorising and action and are central to the decolonial epistemic turn in which the excluded, oppressed, and marginalised are no longer positioned as objects, but are valued as knowers. The turn toward indigenous ways of being, knowing and doing is evident in the development and articulation of local psychologies that centre lived experience and lifeworlds to counteract the coloniality of power. The turn is also evident in the significant intervention of indigenising psychology, and the development of Indigenist paradigms in many contexts across the global South that centre relational ontologies, pluriversality, and situated knowing and relational deconstruction and reconstruction, which is necessary for responding to the multiple crises facing communities across the globe (e.g., Dudgeon & Walker, 2015; Guimarães, 2020; Sonn et al., 2019). Ciofalo (2019), Guimarães (2020), and Dudgeon, Bray, D'Costa, and Walker (2017) offer stimulating additional examples of how indigenous peoples in Mexico, Brazil, and Australia, respectively, are imagining, being, doing, and enacting resistance and self-determination through dialogical processes produced in and through the cultural lifeworlds intertwined within wider histories, structures and experiences.

Critical Mutual Accompaniment, Intersectional Identities, and Dialogical Ethics

Watkins brings another lens to participatory engaged action that is based on relational ethics and critical reflexive praxis in the context of border crossing and work in the contact zone. The author examines the practice of accompaniment from the outside. Martín-Baró, Aron, and Corne (1994) stated:

> the choice is between accompanying or not accompanying the oppressed majorities.... This is not a question of whether to abandon psychology; it is a question of whether psychological knowledge will be placed in the service of constructing a society where the welfare of the few is not built on the wretchedness of the many, where the fulfillment of some does not require that others be deprived, where the interests of the minority do not demand the dehumanization of all (p. 46).

Watkins argues that accompaniment requires a fundamental reorientation for those who have been born into and/or educated into white, economic, and/or social privilege. The focus on power/privilege and the position of outsider requires critical reflection on one's positionality and its effects. Sonn, Baker and Agung's contribution can also be conceptualised as a form of accompaniment. They bring to the foreground arts and community arts practice within a participatory and liberation-oriented framework to support and promote epistemic justice for young people from the African-diaspora in Australia. The work centres on the experiences of young people, their complex personhood, and embodied subjectivities. It resonates strongly with others who have suggested that the work of young people often involves decolonial performances "…of self and subjectivity [that] are, among other things, rituals that seek to keep the body open as a continued source of questions, as a bridge to connect to others, and as prepared to act" (Maldonado-Torres, 2016, p. 27).

The ideas of accompaniment and accountability are central to epistemic justice and the decolonial project. Maldonado-Torres (2016) noted that "[b]uilding the world of the *you* entails an opposition to metaphysical catastrophe, the paradigm of war, and ontological separation" (p. 21). An ontology and epistemology of relatedness drives that decolonial agenda and has been articulated in community and liberation psychology. Such an episteme has been articulated by liberation scholars in Latin America (e.g., Dussel, 2003; Montero, 2007) and is also the basis of the meaning and values systems of Indigenous peoples around the world (Ciofalo, 2019; Rua et al. this volume). This deep orientation that emphasises relationality and dialogue, pushes away from a monological worldview that favours an oppositional ontology and human-centered thinking, toward relational ontologies that decentre and disrupt anthropocentrism (Walsh, 2015; see also Moreton-Robinson, 2004). The relational orientation has been described as central to liberation-oriented approaches that emphasize dialogue, which entails "… another person and Other, who needs to be accepted not as a subject, but as a social actor, who must be respected, who constructs knowledge, who has history, so there must be mutual respect" (Montero & Sonn, 2009, p. 2). Canham et al. similarly emphasise the relationality and ethicality of decolonisation stating that "[i]t is not just a process of eradication—it is about getting to the relational norms of being" (p. 209). Importantly, all of the chapters emphasise a transformed relationship between the knower and known within a historically-informed situated praxis: "As such, it places the theorizer in the midst of people and in a historical, peopled, subjective/intersubjective understanding of the oppressing/resisting relation at the intersection of complex systems of oppression" (Lugones, 2010, p. 747). Praxis from below, and with, restores the traditional subject as epistemic agent.

Conclusion

Bulhan (1985) wrote that "… one essential precondition for a psychology of liberation is to reformulate the psychological outlook, basic human needs, and foci of change in ways that take into account the priorities and frustrations of the oppressed" (p. 256). Since his original cutting-edge work, we have seen the articulation of the decolonial turn and proliferation of writings across disciplines engaging with this turn. Community psychologies in different contexts have begun to embrace this turn in different ways, accentuated by onto-epistemic disruption as a turn away from Cartesian psychologies, and a turn toward collectivist, contextually immersed humanist ways of doing, knowing, and being. It is by no means an easy task to enact disruption and archival reclamation/expansion as a reparative and emancipatory praxis. As noted in this volume and the other volumes in this series, the decolonial turn offers us ways to engage with the archive, to delink, and unthink, and move beyond solidarities to open the borderlands to a new archive of experience and practices—always and constantly in the making and in response to the invocation: to whom are we accountable?

We concur with Ndlovu Gatsheni (2020) that the decolonial project is one that entails planetary liberation dealing with the questions of life chances, life futures, and rehumanizing, particularly for those people who have been degraded into the zone of non-being. Bringing to foreground Fanon's insights, Maldonado-Torres (this volume) also points to the need to unlearn from our disciplines, to abandon and rupture from epistemes that elide or reproduce catastrophe, to learn from various social actors who reveal the magnitude of catastrophic aspects of our world, and to centre re-humanisation and ways of knowing and understanding that prioritises the "cultivation of relationships" (p. xiv). For us, this is a call to reinvigorate and reconnect, collectively and across transnational contexts, to mobilize our glocal community psychologies towards the unfinished project that seeks to "build the world of [the] you" against ontological separation and "pursued with love, as the positive attitude of the damné, and rage, as a form of negation that is inspired and oriented by the positive attitude of love" (Maldonado-Torres, 2016, p. 23). This is by no means easy work, but to be a key contributor to counter-catastrophic science, community psychologists and other social actors must respond in solidarity, in alliances and coalitions, inside and outside of the university. We need to be engaged with movements and collectives in dislodging and dismantling the colonial matrix of power, and creating futures based on embracing conviviality, decolonial love, justice, and cooperation. As is evident in this volume the foundations to make this shift is there to rejuvenate the justice- and liberation-oriented roots of critical community psychologies and chart a course toward decolonial psychologies. The various contributors to this volume have shown some co-ordinates towards such a counter-catastrophic science and transformative social practice.

References

Adams, G., Dobles, I., Gómez, L. H., Kurtiş, T., & Molina, L. E. (2015). Decolonizing psychological science: Introduction to the special thematic section. *Journal of Social and Political Psychology, 3*, 213–238.

Anzaldúa, G. (1999). *Borderland/La Frontera*. San Francisco: Spinsters/Ann Lute.

Bauman, Z. (2012). Times of interregnum. *Ethics and Global Politics, 5*(1), 49–56. https://doi.org/10.3402/egp.v5i1.17200

Bhatia, S. (2018). *Decolonizing psychology: Globalization, social justice, and Indian youth identities*. Oxford, UK: Oxford University Press.
Boonzaier, F., & van Niekerk, T. (Eds.). (2019). *Decolonial feminist community psychology*. New York: Springer.
Bulhan, H. A. (1985). *Frantz Fanon and the psychology of oppression*. London: Plenum Press.
Burton, M., & Kagan, C. (2005). Liberation social psychology: Learning from Latin America. *Journal of Community & Applied Social Psychology, 15*(1), 63–78. https://doi.org/10.1002/casp.786
Carolissen, R. L., & Duckett, P. S. (2018). Teaching toward decoloniality in community psychology and allied disciplines: Editorial introduction. *American Journal of Community Psychology, 62*(3–4), 241–249. https://doi.org/10.1002/ajcp.12297
Ciofalo, N. (2019). *Indigenous psychologies in an era of decolonization*. New York: Springer.
Coimbra, J. L., Duckett, P., Fryer, D., Makkawi, I., Menezes, I., Seedat, M., et al. (2012). Rethinking community psychology: Critical insights. *The Australian Community Psychologist, 42*(2), 135–142.
De Sousa Santos, B. (2007). Beyond Abyssal Thinking: From Global Lines to Ecologies of Knowledges. Review (Fernand Braudel Center), 30(1), 45–89. Retrieved May 18, 2021, from http://www.jstor.org/stable/40241677
de Sousa Santos, B. (2016). *Epistemologies of the South: Justice against epistemicide*. New York: Routledge.
Dudgeon, P., Bray, A., D'Costa, B., & Walker, R. (2017). Decolonising psychology: Validating social and emotional wellbeing. *Australian Psychologist, 52*(4), 316–325. https://doi.org/10.1111/ap.12294
Dudgeon, P., & Walker, R. (2015). Decolonising Australian psychology: Discourses, strategies, and practice. *Journal of Social and Political Psychology, 3*(1), 276–297. https://doi.org/10.5964/jspp.v3i1.126
Dussel, E. (2003). *Philosophy of liberation*. Eugene, OR: Wipf and Stock.
Fernández, J. S., Sonn, C., & Carolissen, R. (2019). Mapping in ad for decolonial transnational critical community psychology. *The Community Psychologist, 52*(4), 4–7. Retrieved from https://www.scra27.org/publications/tcp/tcp-past-issues
Fernández, J. S., Sonn, C. C., Carolissen, R., Stevens, G. (2021). Roots and Routes Toward Decoloniality Within and Outside Psychology Praxis. Review of General Psychology. https://doi.org/10.1177/10892680211002437
Fine, M. (2015). Global provocations: Critical reflections on community based research and intervention designed at the intersections of global dynamics and local cultures. *Community Psychology in Global Perspective, 1*(1), 5–15.
Fricker, M. (2007). *Epistemic injustice*. Oxford, UK: Oxford University Press. https://doi.org/10.1093/acprof:oso/9780198237907.001.0001
Fukuyama, F. (2006). *The end of history and the last man* (1st Free Press trade pbk ed.). New York: Free Press.
Gramsci, A. (1971). *Selections from the prison notebooks of Antonio Gramsci*. New York: International Publishers.
Guimarães, D. (2020). *Amerindian paths through recurrent sociocultural crises*. Retrieved from https://youtu.be/gwNdRcOQ8UY
Humphrys, E. (2015). *Anti-politics and the illusions of neoliberalism. Progress in political economy*. Retrieved from https://www.ppesydney.net/anti-politics-and-the-illusions-of-neoliberalism/
Kagan, C., Burton, M., Duckett, P., Lawthom, R., & Siddiquee, A. (2019). *Critical community psychology: Critical action and social change*. New York: Routledge.
Keeley, B. (2015). *Income inequality: The gap between rich and poor* (OECD insights). Paris: OECD Publishing. https://doi.org/10.1787/9789264246010-en
Kessi, S. (2017). Community social psychologies for decoloniality: An African perspective on epistemic justice in higher education. *South African Journal of Psychology, 47*(4), 506–516. https://doi.org/10.1177/0081246317737917

Kessi, S., Suffla, S., & Seedat, M. (2021). Decolonial enactments in community psychology. Springer New York. https://www.springer.com/gp/book/9783030752002#otherversion=9783030752019

Lugones, M. (2003). *Pilgrimages/peregrinajes: Theorizing coalition against multiple oppressions*. Lanham, MD: Rowman & Littlefield Publishers.

Lugones, M. (2010). Toward a decolonial feminism. *Hypatia, 25*(4), 742–759. https://doi.org/10.1111/j.1527-2001.2010.01137.x

Mair, P. (2013). *Ruling the void: The hollowing of Western democracy*. London: Verso.

Maldonado-Torres, N. (2011). Thinking through the decolonial turn: Post-continental interventions in theory, philosophy, and critique – An introduction. *Transmodernity Journal of Peripheral Cultural Production of the Luso-Hispanic World, 1*(2).

Maldonado-Torres, N. (2016). *Outline of ten theses on coloniality and decoloniality*. Fondation Frantz Fanon. Retrieved from https://fondation-frantzfanon.com/wp-content/uploads/2018/10/maldonado-torres_outline_of_ten_theses-10.23.16.pdf

Martin, K., & Mirraboopa, B. (2003). Ways of knowing, being and doing: A theoretical framework and methods for indigenous and indigenist research. *Journal of Australian Studies, 27*(76), 203–214.

Martín-Baró, I., Aron, A., & Corne, S. (1994). *Writings for a liberation psychology (2. print)*. Boston: Harvard University Press.

Mignolo, W. D., & Walsh, C. E. (2018). *On decoloniality: Concepts, analytics, praxis*. Durham, NC: Duke University Press.

Montero, M. (2007). The political psychology of liberation: From politics to ethics and back. *Political Psychology, 28*(5), 517–533.

Montero, M., & Sonn, C. C. (2009). About liberation and psychology: An introduction. In M. Montero & C. Sonn (Eds.), *Psychology of liberation: Theory and applications* (pp. 1–10). New York: Springer. https://doi.org/10.1007/978-0-387-85784-8

Montero, M., Sonn, C. C., & Burton, M. (2017). Community psychology and liberation psychology. In M. A. Bond, I. Serrano-García, & C. B. Keys (Eds.), *APA handbook of community psychology* (Vol. 1: Theoretical foundations, core concepts, and emerging challenges, pp. 149–167). Washington, DC: American Psychological Association.

Moreton-Robinson, A. (Ed.). (2004). Whitening race: Essays in social and cultural criticism. Aboriginal Studies Press.

Ndlovu Gatsheni, S. (2020, April). Planetary human entanglements and the crisis of living together (Video file). YouTube. https://youtu.be/LgHxsXIMNRM

Ortiz Torres, B. (2020). Decoloniality and community-psychology practice in Puerto Rico: Autonomous organising (autogestión) and self-determination. *International Review of Psychiatry, 32*(4), 359–364. https://doi.org/10.1080/09540261.2020.1761776

Pickett, K. E., & Wilkinson, R. G. (2015). Income inequality and health: A causal review. *Social Science & Medicine, 128*, 316–326. https://doi.org/10.1016/j.socscimed.2014.12.031

Quijano, A. (2000). Coloniality of power and Eurocentrism in Latin America. *International Sociology, 15*(2), 215–232.

Ratele, K. (2019). *The world looks like this from here: Thoughts on African psychology*. Johannesburg, South Africa: Wits University Press.

Ratele, K., Cornell, J., Dlamini, S., Helman, R., Malherbe, N., & Titi, N. (2018). Some basic questions about (a) decolonizing Africa (n)-centred psychology considered. *South Africa Journal of Psychology, 48*(3), 331–342.

Rappaport, J. (1977). Community psychology: Values, research, and action. Holt, Rinehart and Winston.

Reyes Cruz, M., & Sonn, C. (2015). (De)colonizing culture in community psychology: Reflections from critical social science. In R. Goodman & P. Gorski (Eds.), *Decolonizing "multicultural" counseling through social justice* (pp. 127–146). New York: Springer. https://doi.org/10.1007/978-1-4939-1283-4

Rigney, L. (1997). Internalisation of an indigenous anti-colonial cultural critique of research methodologies: A guide to indigenous research methodologies and its principles. *Journal of American Studies, 14*(2), 109–122.
Roy, A. (2020, April 2). The pandemic is a portal. *Financial Times*. Retrieved from https://www.ft.com/content/10d8f5e8-74eb-11ea-95fe-fcd274e920ca
Seedat, M., Suffla, S., & Christie, D. J. (2017). Emancipatory and participatory methodologies in peace, critical, and community psychology. https://doi.org/10.1007/978-3-319-63489-0
Sloan, T. (2018, July). *On waking up to coloniality: An invitation*. Retrieved from http://www.parkmoreinstitute.org/on-waking-up-to-coloniality-an-invitation/
Smith, L. T. (1999). Decolonizing methodologies: Research and indigenous peoples (1st edition). Zed Books.
Sonn, C., Rua, M., & Quayle, A. (2019). Decolonising applied social psychology: Culture, indigeneity and coloniality. In K. O'Doherty & D. Hodgetts (Eds.), *The SAGE handbook of applied social psychology* (pp. 39–57). Boston: SAGE.
Stevens, G. (2016). *Rethinking the psychosociality of violence: Economies of morality, embodied enactments and affective practices*. Inaugural lecture presented at the University of the Witwatersrand, Johannesburg, South Africa, 30 August 2016.
Stevens, G. (2017). Critical psychosocial mnemonics as a decolonising participatory method: Towards reclaiming and refiguring the archive through memory, stories and narratives. In M. Seedat, S. Suffla, & D. J. Christie (Eds.), *Peace psychology book series. Emancipatory and participatory methodologies in peace, critical, and community psychology* (pp. 179–199). New York: Springer. https://doi.org/10.1007/978-3-319-63489-0_14
Stevens, G. (2018a). Raced repetition: Perpetual paralysis or paradoxical promise. *Keynote presentation at the (Re)imagining Liberations: Instutionalised Despair*Critical Hope Conference,* 06–08 August 2018. Johannesburg: University of the Witwatersrand.
Stevens, G. (2018b). Raced repetition: Perpetual paralysis or paradoxical promise. *International Journal of Critical Diversity Studies, 1*(2), 42–57.
Stevens, G., Duncan, N., & Sonn, C. (2010). The apartheid archive: Memory, voice and narrative as liberatory praxis. *Psychology in Society, 40*, 08–28. Retrieved May 31, 2020, from http://www.scielo.org.za/scielo.php?script=sci_arttext&pid=S1015-60462010000200002&lng=en&tlng=en
Sullivan, S., & Tuana, N. (Eds.). (2007). *Race and epistemologies of ignorance*. Albany, NY: State University of New York Press.
Teo, T. (2005). The critique of psychology: From Kant to postcolonial theory. Springer.
Therborn, G. (2009). The killing fields of inequality. *Soundings,* 42. London: Lawrence and Wishart.
Visvanathan, S. (2009). *597 Shiv Visvanathan, The search for cognitive justice*. Retrieved May 25, 2020, from https://www.india-seminar.com/2009/597/597_shiv_visvanathan.htm
Walsh, C. E. (2015). Decolonial pedagogies walking and asking. Notes to Paulo Freire from AbyaYala. *International Journal of Lifelong Education, 34*(1), 9–21. https://doi.org/10.1080/02601370.2014.991522

Africa's Knowledge Archives, Black Consciousness and Reimagining Community Psychology

Shahnaaz Suffla and Mohamed Seedat

Each generation must out of relative obscurity discover its mission, fulfill it, or betray it.

(Fanon, 1963, p. 206)

The epistemic landscape upon which community psychologies have flourished has come under close scrutiny over the last number of decades, with theoretico-praxical analyses emanating largely from work inspired by humanistic and critical scholarly traditions in South Africa, Latin America, the Middle East, Australia, New Zealand, and some parts of the Global North (see Seedat & Suffla, 2017). A characterising focus of these analyses is the production of knowledge that contests dominant epistemological and ontological claims, and which privileges ethico-political and social justice imperatives central to human well-being in an increasingly oppressive social order. Moreover, the recent decolonial turn in the social sciences and humanities has extended the continuities, but also revealed the disjunctures and omissions in critical thought in community psychology. Accordingly—and in resistance to the inherited Euro-American knowledge archive in community psychology—in this chapter we offer a historicised reading of Africa's knowledge archives as a partial analysis of the archaeology of knowledge on the continent and its confluences with community psychologies. We extend this analysis to consider Black Consciousness philosophy more specifically for its relation to, and possibilities for, a decolonial community psychology. As an enactment of intellectual insurgency against

S. Suffla (✉) · M. Seedat
Institute for Social and Health Sciences, University of South Africa, Johannesburg, South Africa

South African Medical Research Council-University of South Africa Masculinity and Health Research Unit, Cape Town, South Africa
e-mail: ssuffla@mrc.ac.za

G. Stevens, C. C. Sonn (eds.), *Decoloniality and Epistemic Justice in Contemporary Community Psychology*, Community Psychology,
https://doi.org/10.1007/978-3-030-72220-3_2

coloniality, and in accord with Luis Macas's (2004) pronouncement that "the values, principles, knowledge, wisdom of our peoples must not only be recovered and archived" (para. 5), we present this chapter as "a contribution from our peoples to society as a whole as a function of change, as substantial elements of an alternative approach" (Macas, 2004, para. 5).

We are led by Stuart Hall's (2001, p. 92) argument that:

> Archives are not inert historical collections. They always stand in an active, dialogic, relation to the questions which the present puts to the past; and the present always puts its questions differently from one generation to another.

It is in this vein that we strive here to re-centre certain African knowledge archives so that we may consider the reconstruction possibilities for community psychology evident from archival engagements, and that uphold a decolonial view of humanity (see Maldonado-Torres, 2017). In so doing, we seek to advance the decolonising project in community psychology in Africa. We draw on select historiographic accounts of Africa (e.g., Mazrui, 2005; Mudimbe, 1988; Zeleza, 2007) to invoke, and reiterate, Africa's multiple archives and their attendant paradigmatic traditions. Our engagements with these archives signify at once *onto-epistemic rupture* as a counterpoint to the Eurocentric assumptions that continue to frame community psychology theory and practice in Africa and other parts of the Global South, and *epistemic praxis* as a decolonising response, inspired by the subalternised knowledges of Africa. As such, the notional foundations of community psychology on the continent might well be reorganised. Indeed, the quest for decolonised community psychologies must necessarily engage critically with multiple knowledge archives and their conjunctions. From this perspective—and given the turn to the archive as an account of knowledge arrangements, and as a transdisciplinary analytic—trans-archival critical conversations may yield yet unexplored and unelaborated theoretical and pragmatic spaces for decolonising community psychologies. Thus, we consider a continuing interpretation of archival accounts as a requisite interjection (in)to the recognition, recuperation, (re)assembly, remaking, and repositioning of such histories in the present. Indeed, we believe it is what the present compels us to do. Within this process, we argue that the intellectual extension of Black Consciousness, as articulated by Bantu Stephen (Steve) Biko and others in South Africa, an Africa-inspired knowledge archive produced in philosophical, social and material context, is a particularly seminal onto-epistemological and methodological contribution to Africa's knowledge archive.

It is also perhaps important to note that the process of engaging with these multiple archives represents for us a reflexive, philosophical, and affective encounter that is as evident and evidenced as it is obscure and obscured. As such, our desire is to write "with and against the archive," as Hartman (2007) so incisively observes.

African Archives: Representations, Rupture and Retrieval

> *... and brings them into contact with the analyst's mind, thus inventing in a dynamic manner both understanding and history*
> *(Mudimbe, 1988, p. 212)*

Zeleza (2007) records that several intellectual traditions materialised across four epochs of African historiography, representing the frontline between Afrocentricm and Eurocentricism. While African historiography is of course much older than disciplinary accounts, other constructions of African intellectual practices exist (e.g., see Mazrui, 2005), and there are some that are entirely specific to countries on the continent (e.g., Vale, Hamilton, & Prinsloo, 2014), in this chapter, we lean primarily on Zeleza's (2007) historiographical account of African knowledge libraries. The first period of this historiography, constituting the pre-colonial era, produced the Afro-Christian, Afro-Islamic and griot archives. Espoused by theologists and philosophers, and in particular through the writings of St Augustine from present-day Algeria, the Afro-Christian archive centred divinity and perceived history as universal to the extent that it involves movement towards divine providence. Assuming Christian normative foundations, this archive extended to Ethiopia, where knowledge was produced to proclaim dynastic grandeur through the grammar of religion. The Afro-Islamic archive, which emerged from the expansion of the Islamic frontier in Africa, advanced Islamic theology and philosophy, and encouraged the dispersion of Greek, Roman and Asian intellectual ideas to Western Europe. This archive is strongly influenced by the contributions of Ibn Khaldun from present-day Tunisia, whose work challenged the notion of providential history and instead privileged a cyclical theory of history. The Islamic tradition inspired the establishment of many centres of learning in Africa, including Al-Azhar University in Egypt, University of al-Qarawiyyin in Morocco, and the scholarly academy of Timbuktu, all representing significant repositories of knowledge. The Timbuktu archive is considered as a generative centre of Afro-Islamic scholarship that integrated studies of religion and science, challenging the idea that heritage preservation occurred only through the oral tradition and in European centres of knowledge (Goolam, 2006). Mazrui (2005) characterises the Timbuktu archive as "black", as a site that catalysed black consciousness in Africa, and—in interaction with the universities—as embodying black scholarly excellence. The griot archive was founded on oral and narrative traditions, which aimed to connect the past and the present, build collective cosmologies and identities, articulate political opinions, instruct youth in the ways of knowing, and provide aesthetic pleasure. The griots, as skilled custodians of oral and narrative traditions, undertook multiple roles in their societies, such as serving as diplomats, teachers, warriors, political advisors, mediators and entertainers.

Collectively, the Afro-Christian, Afro-Islamic and griot archives honoured multiple epistemological and ontological positions. These archives emphasised mutual inter-dependence, syncretic relations between nature, people and society, and relationality where beingness is affirmed through the inter-subjective context (see Dei, 2000). From the perspective of these African systems of thought, reality is comprehended through an integrated view of society, where knowledge is situated in social experience and reality, and is collective in its genealogy. These archives perceived modes of being as inherent to forms of knowledge-making that assume psycho-spiritual and ethical praxes. Hence, when considered in relation to psychology broadly, and community psychology specifically, these knowledge archives would have produced particular understandings of human nature and human behaviour, underlined by notions of syncreticity, the collective and psycho-social well-being. Notwithstanding, these archives are clearly open to further interpretation for their import to the psychologies of Africa.

The second period, spanning from the fifteenth to nineteenth centuries, and associated with the transatlantic slave trade, produced the Eurocentric and vindicationist traditions (Zeleza, 2007), representing the colonial and counter-colonial archives respectively. The Eurocentric tradition constructed Africans as primitive, infantile, savage-like, lacking in agency and reason, and as located outside the zone of humanity. Africa was manufactured as static, ahistorical, and as immured in monolithic modes of living (also see Mudimbe, 1988; Ndlovu-Gatsheni, 2013). Both Africa and Africans were cast as lacking historicity (and futuricity). In contrast, the counter-archive of this period created the impetus for a nationalist and Pan-Africanist philosophy, and historiography, mostly during the twentieth century. This archive validated Africa's humanity and historical ontology, as well as recorded its successes and contributions to civilisation.

The third period, emerging in the twentieth century, naturalised and entrenched the colonial archive that arose in the preceding period. Through the register of modernity, this archive expunged African knowledges through omission, denigration, and distortion. However, as Zeleza (2007) points out, this colonial archive did not go unchallenged: imperialist perspectives were countered by nationalist and radical outlooks that chronicled Africa's history as going back millennia where the former affixed it to the arrival of the colonisers. The contest of the colonial knowledge project unfolded here in the mold of the vindicationist tradition to produce nationalist historiography that centred African subjectivity, agency, self-determination and the struggle against colonialism. This early iteration of decolonisation on the continent was accompanied by the establishment of new universities and a new wave of scholarly activities.

The colonial archives of the second and third periods introduced dominating knowledge systems which subverted other-than-European intellectual cartographies and imposed European constructs, denoting problematic forms of knowledge-making on African traditions, cultures and polities. Founded on the direct control of colonised people's lands and psyches, colonialism manufactured knowledge through the exploitation of 'dependencies' (Mudimbe, 1988), to the advantage of the colonial economic system. Aligned with Europe's politico-economic objectives, which privileged the industrial revolution, heavy industry, tertiary and service related work, and exports, the onto-epistemological structure that authorised the colonising project cultivated the basis for the othering and discrediting of colonised societies, cultures and people, thereby effecting a dichotomising system of knowledge. This dichotomising system naturalised supposed differences between the modern and traditional, the oral, written and printed, and the Occidental, Oriental and African.

This colonising structure that naturalised othering was buttressed by missionaries' discourses on the colonised and the theoretical claims of scientific racism, which assumed the logic of organising human beings and societies along a hierarchical continuum of increasing developmental superiority. Missionary and racial scientific discourses, along with European explorers' narratives, constituted three genres of "speech" that propagandised "the invention of a primitive Africa" (Mudimbe, 1988. p. 82). These modes of speech were manifest in the 'exotic' texts produced by European travelers in the late seventeenth century, the eighteenth century philosophical accounts of a hierarchy of civilisations, and the nineteenth century anthropological pursuits for primitiveness, or what may be more rightly viewed as the search for 'the primitive'. The three genres of speech formed the intellectual practices for negating, reducing and inferiorising other-than Western civilisations, societies and people (Mudimbe, 1988; also see Ndlovu-Gatsheni, 2013; Ngũgĩ wa Thiong'o, 1986).

Within this context, the social sciences, including psychology, was highly complicit. Psychology, integral to the colonial knowledge systems, conducted research in and on Africa that was framed by ideologically-loaded instrumental reasoning. Africans were both the subject and object of psychological research and theorising that was employed to make Africans' exploitation effective. Just as Africa's human and material resources were historically (and continue to be) exploited to develop Europe's technology and institutions, Africans were constructed as such to promote the development of Euro-American psychology and its conceptions of human nature and non-being (Bulhan, 1981; Rodney, 1972). For instance, anthropology and psychology proposed what MacGaffey (1981, p. 262) calls "the mythological phase", namely the formulation of hypotheses that juxtaposed the idealised qualities of European civilisation with their assumed absence in Africa. Irrespective of their variabilities, missionary, psychological and anthropological discourses explicitly shared a belief in the 'modifiability' of African culture and the view that colonisation is both a "providential accident and a historical benefit" (Mudimbe, 1988, p. 80). The process of primitivisation and othering contained two substantial consequences. First, it provided the intellectual justification and rationalisation for the dispossession, subjugation and occupation of Africa. Whereas colonial soldiers secured the violent occupation and suppression of Africa, psychology's body of knowledge—of course, among others—offered the scientific and disciplinary justifications for restraining the consciousness and conscience of the administrators and soldiers of colonialism. Second, as primitivisation and occupation were increasingly legitimated and authorised in scientific discourse, the colonialists and missionaries accrued validation to continue with their evangelising passage to convert Africa(ns) to civilisation and idealised western modernity (Bulhan, 1981; Mudimbe, 1988). In turn, these two processes operated recursively to lend justificatory support to each other.

Framed by the colonial archive of the early twentieth century, South African social-community interventions, which concentrated on the social, economic and industrial challenges faced by the minority white population, endorsed the claims of the colonial knowledge structure in the country. These interventions, which contributed greatly to the authorisation and professionalisation of Eurocentric psychology in the country, provided pseudo-scientific rationalisations for segregationist social planning, and exclusionary policy formulations and practices, including the construction and defense of privileged racial communities (Louw, 1986a; Nicholas & Cooper, 1990; van Ommen & Painter, 2008). Although not formally denoted as such, the First Carnegie Commission Study (Carnegie Commission Report, 1932) represented one instance of a large-scale social-community intervention that focused on the mitigation of poverty among the white working class, triggered by anxiety about the socio-political consequences of emerging class divisions in the White Afrikaner social order (Louw, 1986a, 1986b). Along with educationists and other experts, psychologists participating in the Carnegie Study proffered that, in the context of white poverty, class distinctions were dissolving an assumed sense of community among whites. Instructed by worries about the decline of an assumed monolithic whiteness, fears of a black peril, and the intensifying social interaction between poor whites and Africans, heralding working-class solidarity (Carnegie Commission Report, 1932; Carton, 1990; Louw, 1986a; Seedat, 1993, 1998), the psychologists' submissions to the Carnegie Study (Carnegie Commission Report, 1932) sanctioned the government's policy of workplace and social segregation. Notwithstanding its engagement of civil society, and the emerging social and psychological sciences, the First Carnegie Study's racialised and racist accounts of poverty and related proposals represented a disciplinary intervention that sought to authorise and preserve white privilege and the making of racially segregated communities as a response to increasing inter-racial working-class solidarity (see Appel, 1989; Carnegie Commission Report, 1932; Louw, 1986a; Seedat, 1993; Seedat & MacKenzie, 2008). Thus, psychology in South Africa was imbedded from its very incipiency in the epistemic economies and violences of the ideological, discursive and material racism that constituted/s the colonial archive, where the standpoint of domination (see Almeida, 2015) propagated and legitimated white supremacy, while infantilising, inferiorising and invalidating the capacities of black South Africans to know, represent, and act for themselves and their communities.

Returning to Zeleza (2007), the fourth period, animated from the 1970s onwards, represents the post-colonial era. At this time, tensions surfaced between the vindicationist nationalist and radical traditions noted earlier, with nationalist historiography criticised for the lack of systematic enquiries into the formations and sequelae of imperialism. The radical tradition, which encompassed Marxist, dependency, feminist, ecological and postcolonial thought, adopted diverse conceptual constructions. These included a respective focus on: the class struggle, labour and workers, and the shifting configurations of Africa's economy; Africa's 'under-development', and its external dependency for capital and resources; the (dis)placement of women in African historiography, and the hegemony of western feminism; the multi-faceted interactions between social ecology, nature, history and geography, and the interaction between power and the environment; and analyses that—in the 'posts' tradition—privilege discourse in the understanding of historical experience, and attend to such notions as hybridity, contingency and ambivalence (see Zeleza, 2007). Despite its non-homogeneity, the radical tradition gave rise to an array of epistemes that have been incorporated into Africa's historiography.

Perhaps in some respects reflective of the multiple libraries that characterised the third and fourth periods of African historiography, and pushing back against hegemonic articulations, South Africa witnessed various expressions of community psychology during the 1980s and 1990s. These early manifestations of community psychology—as they embodied a counter response from intellectual communities within the discipline—were reflective of liberal as well as black, feminist, critical, indigenous, liberation and Marxist perspectives. Against the backdrop of the struggle for national democracy, the inclination towards community psychology in South Africa was most evident in the scholarly contributions that attempted to address the onto-epistemological and material manifestations of "the epistemic erasures, omissions, fabrications, stereotypes and silences" (Zeleza, 2007, p. 15) on which the colonial archive gained its ascendancy, and to which the discipline of psychology remained resolutely tethered. This pluriversal and disruptive counter archive—and one variant—in community psychology is well documented and referenced (see Seedat & Suffla, 2017 for overview). While we do not have the space to describe fully this archive here, in the composite, we consider this body of knowledge in South African psychology—in part reflective of the radical traditions of the fourth period of Africa's historiography—to be an axiological contribution towards archival resources supportive of transformation and social justice aspirations, collective welfare and liberation-leaning theories. This is not to say, of course, that all of community psychology in South Africa is representative of this oppositional archive. Its disciplinary derivations, as described here, are—in themselves—certainly not a patent signifier of community psychology's liberatory ideals and praxes. Indeed, community psychology's locations alongside, outside and within the dominant psychologies of the twenty-first century, and in relation to the modern state—characterised by the centralisation of power, authority and control—inexorably and precariously places it between the forces of complicity and cooption, and resistance and change.

We wish, however, to single out the late 1980s-early 1990s contributions of the Psychology and Apartheid Committee as a conjunctural moment in our argument, and as a prologue to the section that follows. The Psychology and Apartheid Committee, emblematic of critical black perspectives and independent black intellectual thought (Seedat, 1993; Yen, 2007, 2008), and drawing inspiration from the decolonial works of Biko (2004), Bulhan (1985, 2015), Cabral (1979), Fanon (1963, 1965, 1967), Ngũgĩ wa Thiong'o (1986), Shariati (1974) and other anti-colonial scholars, centred the black experience of apartheid-colonialism and slated psychology for ignoring the perniciousness of apartheid and its impacts on the majority population's psycho-social well-being (Cooper, Nicholas, Seedat, & Statman, 1990; Letlaka-Rennert, 1990; Nicholas, 1991; Nicholas & Cooper, 1990; Seedat, 1993). Representing decolonising voices, the Psychology and Apartheid Committee located collective psychological liberation within the broader struggles for democracy, and insisted on including blacks and women in the processes of knowledge creation, training, research and organised psychology (see Nicholas & Cooper, 1990; Seedat, 1993). Resonant with the contemporary iteration of the decolonial turn, that "generates a decolonial attitude that takes primacy over method, allowing subjects to uncover fundamental problems and to join others in the efforts to address them" (Maldonado-Torres, 2018, p. 125), the Committee's work was an attempt at making oppression visible in relation to knowledge, power and being, and foregrounding and authenticating psychological liberation as a subject focus area for community psychology and other fields of psychology. The decolonising contributions of the Committee generated a liberatory attitude that reimagined psychologies, and specifically community psychologies, for Africa, and inspired aspirations of epistemic, distributive and representational justice. This first wave of black-led decolonisation provided pathways to revisit and influence the intellectual trajectory of community psychology in Africa within a more historically, or archivally-engaged, knowledge context.

The contemporary decolonial turn in South African psychology, prompted most notably by African, Latin American and Caribbean liberation thought and philosophies (e.g., de Sousa Santos, 2018; Fanon, 1963, 1965, 1967; Gordon, 2000; Grosfoguel, 2013; Maldonado-Torres, 2007, 2008, 2016; Mamdani, 2012; Mignolo & Walsh, 2018; Ndlovu-Gatsheni, 2013, 2018), has extended the paradigmatic offerings of this earlier period—its pre-history (see Hall, 2001)—to further decentre hegemonic psychology and reorient its construction, content and consumption. To this end, critical scholarship, methodologies and projects in this current iteration of decolonising psychology—inclusive of the contributions of critical community psychologists—are focused on various social inscriptions of power, difference and oppression, such as 'race', gender, class and sexuality; interrogations of research and intervention practice; transformation of teaching, pedagogies and curricula; the university as a site of subjugation and emancipation; and the (de)colonial specificities of the knowledge economy (see Barnes & Siswana, 2018; Seedat & Suffla, 2017). In its striving towards what Dei (2018, p. 135) refers to as a "coming to know", the decolonising project specifically in social-community psychology has resulted in the publication of two Special Issues of the South African Journal of Psychology, the first focused on Liberatory and Critical Voices in Decolonising Community Psychology (see Seedat & Suffla, 2017), and the second on Psychology and Decolonisation (Barnes & Siswana, 2018); the creation of the Hub for Decolonial Feminist Psychology in Africa (see http://www.humanities.uct.ac.za/news/uct-launches-hub-decolonial-feminist-psychologies-africa); the launch of the Transdisciplinary African Psychologies knowledge platform (see https://www.unisa.ac.za/sites/corporate/default/Colleges/Graduate-Studies/Schools,-Institutes-&-Research-Chairs/School-of-Transdisciplinary-Research-Institutes-(STRI)/Institute-for-Social-and-Health-Studies-(ISHS)/Transdisciplinary-African-Psychologies); and a mini-series on decolonising community psychology in the Springer Community Psychology Book Series (inclusive of the current text), under the editorship of South African and other Global South colleagues (see https://www.springer.com/series/15965 for Series description). These initiatives represent exemplars of oppositional knowledge production within the discipline, iterative of similar work underway in the greater Global South, to which Africa-located scholars have also contributed (e.g., Carolissen & Duckett, 2018; Tuğçe, Gómez, Molina, Dobles, & Adams, 2015). To a large extent, the politicised intellectualism, praxes, collectives and relations that constitute this evolving African liberatory archive in psychology historicises and centres blackness and anti-blackness both in the interpretation of the logics of coloniality, as well as the articulation of decolonising imaginaries. It is against this framework that we theorise Black Consciousness philosophy as a distinctive and insurrectionist archive that may be mobilised and engaged for liberatory articulations of community psychology in Africa, and equally as a challenge and potential mode of reconstruction for community psychology.

Black Consciousness as an Insurgent Archive in Community Psychology

> *Black Consciousness unequivocally emphasized the need for psychological liberation (AZAPO, n.d., n.p.)*

The Black Consciousness Movement emerged in the 1970s to become one of the most influential political formations in apartheid South Africa (see Mngxitama, Alexander, & Gibson, 2008). Its foundations were laid by the South African Students Organisation (SASO), which was officially launched in 1969, with Steve Biko as its leading thinker. Influenced by decolonial theorists such as Frantz Fanon, Julius Nyerere, Aimé Césaire and William Edward Burghardt Du Bois, the Movement sought to foster black consciousness and African consciousness towards restoring the humanity, and realising the physical and psychological liberation, of black South Africans. This consciousness was envisioned as a potent and solidarity-driven counter to the profound and assiduous offensives and insults of colonialism and apartheid on the black psyche. With "black man (*sic*), you are on your own" as its rallying cry, Black Consciousness philosophy asserted positive and politicised constructions of black identity, and black pride as non-contingent on white recognition. Importantly, it located change within the subjectivity and locus of the oppressed. More (2008, p. 50) explains that Black Consciousness:

> was and still is a struggle for a new consciousness, a reawakening of a self-consciousness, a re-appropriation of black self consciousness from the clutches of an appropriative and dominating white consciousness, a rediscovery of the black self which lay buried beneath white consciousness imposed on blacks by ... domination.

A substantial body of scholarship exists—for example, in political science, philosophy, history, and literary and visual arts literature—that has been inspired by the Black Consciousness Movement, and the intellectual and political contributions of Steve Biko (see Hadfield, 2017). This work demonstrates that Black Consciousness philosophy remains as relevant today as it was in the previous century, particularly in the face of the ongoing struggles for material, psychological, and epistemic emancipation in South Africa, but elsewhere too. A more recent occasion of the influence and mobilisation of Black Consciousness philosophy is perhaps best illustrated by the "Rhodes Must Fall" and "Fees Must Fall" student movements. Reawakened in 2015 to protest racism at South African universities, and increases in student fees, the students' activism recalled historical concerns about the coloniality of power, knowledge and being, and the coloniality-social sciences nexus. In psychology specifically, critical South African contributions through the theoretical prism of Black Consciousness include work by Manganyi (1973), Hook (2004), Seedat and Lazarus (2014), and Cooper and Ratele (2018), who—through somewhat different inflections—collectively call for the inclusion of radical psycho-political praxes in psychology. Our chapter builds on their provocations. In particular, we find impetus in Seedat and Lazarus' (2014) formulation of Black Consciousness grounded community programmes as the "precursor to the formalized radical forms of community psychology in South Africa" (p. 270), as well as Cooper and Ratele's (2018) reference to Biko's and Black Consciousness philosophy's "focus on engagement with people in their lived experiences, with renowned self-reliance programmes" (p. 251), where "this dual focus on the black self and group" (pp. 251–252) "seeks to liberate the self and the group" (p. 252). We also seize on Manganyi's (1973) reading of Black Consciousness as "consciousness of past, present and future" (p. 19). For us, then, the noted potentialities and temporality of Black Consciousness, as located within a decolonising imaginary and drawing primarily from the writings of Steve Biko, offer the prospect of an alternative—*insurgent*—archive of knowledge, that remains under-studied in community psychology.

Hall (2001, p. 90) submits that:

> The field of an archive is marked by rupture, significant breaks, transformations, new and unpredicted departures. The trick seems to be not to try to describe it as if it were the oeuvre of a mythical collective subject, but in terms of what sense or regularity we can discover in its very dispersion.

The analysis that we offer below is inclined towards Hall's (2001, p. 91) contention that ultimately it is to transformation that such a "regularity-in-dispersion" perspective is bound. Our analysis is furthermore in appreciation of the sentiment that archives, as the "history of the present", are inevitably connected to "futurity and contingency" (Hall, 2001, p. 92). It is in an extant moment of that futurity that we entreat Black Consciousness as an African archive of meanings and representations that may be mobilised to offer liberatory and historically-located possibilities for the praxical configurations of present-day critical community psychology. Our conceptions—partial and evolving—accord primacy to the exposition of *collectivist and humanising ontology* as a foremost contribution by the Black Consciousness Movement to the making of liberatory enactments of community psychology. We elaborate below.

> The first step therefore is to make the black man come to himself; to pump back life into his empty shell; to infuse him with pride and dignity, to remind him of his complicity in the crime of allowing himself to be misused and therefore letting evil reign supreme in the country of his birth. This is what we mean by an inward-looking process. This is the definition of 'Black Consciousness'. (Biko, 2004, p. 31)

Black consciousness constructs emancipatory ontologies as cardinal "toward[s] a new humanism" (Fanon, 1967, p. 1), where the quest is for liberation and decoloniality (see Maldonado-Torres, 2016) and where "freedom is found with other black people in the struggle for liberation itself" (N. Maldonado-Torres, personal communication, November 17, 2020). Such ontologies, located within the onto-decolonial turn (see Ndlovu-Gatsheni, 2018), expose more unequivocally "modes of being otherwise, those of the oppressed and silenced peoples, peoples that have been radically excluded from the dominant modes of being and knowing" (de Sousa Santos, 2018, pp. 2–3). They eschew the dualisms of psychological and political liberation (the one is unimaginable without the other), and call for re-imagining and restoring generative human relationships with the self and others, and equally with the material and incorporeal (see Ngũgĩ wa Thiong'o, 1986) to counter the compound violence of inferiorisation and dehumanisation. From the vantage point of emancipatory ontologies, and as a contribution to African humanism, the corporeal integrity of the black body in not founded on a separation from consciousness—as is apparent in our elaborations below—but its facticity is instead upheld as an active presence (see Manganyi, 1973; Wilderson, 2008). This presence is constituted of the "profound unity between body and soul, matter and spirit" (Veriava & Naidoo, 2008, p. 237–238), and the "means through which we reinforce ourselves against the violence of European domination" (p. 238) (see Senghor, 1994).

Unlike constrained inscriptions of identity (Ndlovu-Gatsheni, 2013), Biko imagines African subjectivities beyond the signifiers of woundedness and traumatisation. To this end, Black Consciousness strives to infuse and celebrate black pride and black love, that is forms of ontic dignity attached to human- and African-centred notions of self and community where the colonial and apartheid projects equated blackness with non-being. Simultaneously, it calls on the oppressed to recognise the imposition of ontological inferiority by whiteness on the black body—the collective experience of the negation and othering of blackness and the humanity of Africans—where humanity is reduced to an image of whiteness, and whiteness is the epitome of beauty and virtue. Blackness is in this way constructed as a site of agentic subjectivity rather than disavowal, and imbued with the potential for (re)covered and (re)figured formations and enactments of black identity that proclaim ontic presence. Such articulations, focused on black power and African humanism, thus appoint decolonisation of the mind as front and centre in claims for the personhood and humanity of the black person. Here, where conceptions and lived realities of humanity, identity and liberation are at the epicentre of personal, political, intellectual and social labour, "the leitmotif is to constantly raise the question of blackness" (More, 2014, p. 177).

However, in his treatment of questions of identity, Biko has been criticised for his almost exclusive focus on black men and resultant neglect of the gender oppression of black women (see Dineo Gqola, 2001), signaling a seeming paradox, considering that black women's agency in the Black Consciousness Movement was highly visible and extolled (see Moodley, 2018). Conversely, more nuanced and historically-situated scrutiny suggests that the masculinised use of language—as evident in the Biko quotes that are referenced in this chapter—is a pre-politically correct linguistic peculiarity that should not detract from the Movement's attribution of human centredness and black pride to all, the group, and not men alone [e.g., "The philosophy of Black Consciousness, therefore, expresses group pride and the determination by the blacks to rise and attain the envisaged self" (Biko, 2004, p. 74)] (see Mangena, 2008; Oliphant, 2008). Oshadi Mangena (2008) argues that while gender was not an appointed organising principle of the Black Consciousness Movement, its legitimacy was irrefutably endorsed, albeit tacitly. She adds that while the language use of the time may have appeared insensitive to the gender problematic, this unease becomes variable when examining the Movement's attention to gender through the linguo-cultural lens of African languages; for example, in the grammar of isiXhosa, there is absence of pronominal gender markers. Moreover, at the time, the focus on racial liberation is argued to have been the principal priority, prevailing not as an intentional omission of gender as much as a strategic choice of unity and solidarity to fortify resistance towards the apartheid state (see Magaziner, 2011). These considerations suggest that the ontologies of gender are in fact not diminished and exterior to the analyses of collectivist and humanising ontology that we offer here; instead, we are directed to the need for an intertemporal interpretation of gender in Black Consciousness ideology that allows for a critical engagement with gender and sexual politics.

Biko (2004) states that "the interrelationship between the consciousness of the self and the emancipatory programme is of paramount importance" (p. 53). In this schema, interiority and consciousness are represented within a socio-psychic reality that places the African being indelibly in a relational context—after all, consciousness is always relational. Biko's affirmation that "we must seek to restore to the black man the great importance we used to give to human relations" (Biko, 2004, p. 106) is an attempt to restore a relational ontology which exceeds that constructed by the subordinated relation of blackness to whiteness ["While it may be relevant now to talk about black in relation to white, we must not make this our preoccupation, for it can be a negative exercise" (Biko, 2004, p. 108)]. For Biko, this embodies in large part black solidarity and unity as the antithesis to anti-black racism, and perhaps more fundamentally the very condition of possibility for being-in-the-world. As a unifying and humanising ontological orientation, which translates into the "community of being", this relationality is intrinsic to comprehending the "black experience of being in the world" (Manganyi, 1973, p. 41).

This ontology of being-in-relation also encompassed conceptions of collective agentic subjectivities that were to be mobilised to assume a liberatory attitude ["Black Consciousness is an attitude of mind …" (Biko, 2004, p. 101)], manifest in consciousness of oppression and its internalisation. Biko understood collective agentic subjectivities as possessing the generative capacities to at once pursue freedom, expel the internalised oppressor, and celebrate blackness and black authenticity through the "gift" of African conceptions of emancipation and humanisation to the country:

> We have set out on a quest for true humanity, and somewhere on the distant horizon we can see the glittering prize. Let us march forth with courage and determination, drawing strength

from our common plight and our brotherhood. In time we shall be in a position to bestow upon South Africa the greatest gift possible—a more human face. (Biko, 2004, p. 108)

Biko's commitment, epitomised in the mobilisation of the collective "we", for the benefit of all of "South Africa", confers an emancipatory agency and humanising praxis that disallows absolute victimhood to reify blacks to their full potential. Importantly, Biko assembles this vision as an offering of collectivist selfhood to the country (and beyond), contributing ultimately to a liberatory culture of humanisation which seeks to break from colonial discourses that position black subjects as incapable, dependent, and in need of rescue. There exist then, in this revolutionary enactment of humanism, forms of communalism that are axial to modes of being and becoming for the black subject.

Black culture above all implies freedom on our part to innovate without recourse to white values. This innovation is part of the natural development of any culture. A culture is essentially the society's composite answer to the varied problems of life. We are experiencing new problems every day and whatever we do adds to the richness of our cultural heritage as long as it has man as its centre. (Biko, 2004, p. 109)

Perhaps Biko's writing on African cultural concepts stands as the most direct articulation of collectivist African ontologies. He states that "one of the most fundamental aspects of our culture is the importance we attach to Man" (Biko, 2004, p. 45), describing the African as a cultural being, with African culture centring the person. In Black Consciousness then, the mode of being that is shaped by African culture celebrates intersubjective connection, collective well-being, and joint action. The intersubjective space is not merely an instrumentalist space in which to engage, but is one of intimacy, compassion and sharing in which conversations and social interactions are legitimate in their own right. From the perspective of this collectivist ontological register: individuals are perceived to be inherently good and socially inclined ["hence all our action is usually joint community-oriented action rather than the individualism which is the hallmark of the capitalist approach" (Biko, 2004, p. 46)]; music and dance are held as cognitive-affective resources in communication, and the making and mobilising of community towards resistance and, concomitantly, affirming black modes of being ["Music in the African culture features in all emotional states" (Biko, 2004, p. 46)]; and reciprocal engagement and collective ownership are valued as community-based norms ["African society had the village community as its basis" (Biko, 2004. P. 47)]. Interestingly, Biko adds that "another important aspect of the African culture is our mental attitude to problems …" (Biko, 2004, p. 48), where he formulates the approach adopted by Africans as "situation-experiencing", not as a passive life orientation, but an ontological position where "the African personality with its attitude of laying less stress on power and more stress on man is well on the way to solving our confrontation problems" (Biko, 2004, p. 49). Here, Oliphant (2008) cautions against narrow interpretations, which act to endorse the colonial trope that Africans possess distinct personality and thought. The African culture and humanism thesis in Biko's writings is fundamentally in recognition of the actual and threatened destruction of the culture of the oppressed—its social formation—by the colonisers, and equally of the dynamic nature of culture where fundamental features of its collectivist ontology remain and are recoverable. Oliphant (2008) elaborates that this thesis is not a pseudo-scientific analysis of an ethno-cultural group, but a theoretical intervention to "restore the intrinsic value of African culture to Africans", "present it as a basis for inter-African solidarity", and "critically unmask the pretensions of colonial culture" (p. 218).

Countering ontological dualisms that separate the secular from the spiritual, and the individual from their social ecology, Black Consciousness philosophy references spirituality as a key feature of being. From this perspective, subjectivity is enculturated by its encounter with spirituality in as much it is in relation to the self and the collective. Biko's (2004) reflections on the "spiritual poverty of the black people" (p. 34) is an expression of his concern about forms of phenomenological deprivation resulting from colonisation. His thinking appeals to a collectivist ontology that is African located, one that appeals to the relational elements of humanity and notions of being. That is, ontological restoration through the recovery of spirituality summons an African notion of personhood and humanity (*ubuntu*), captured in the profound Zulu phrase *umuntu ngumuntu ngabantu* (a person is a person through other persons). This is in sharp divergence to Cartesian philosophy—"I think, therefore I am"—that constructs a subject as "an unquestioned ideal of self expressed in the notion of the *ego conquiro* [I conquer]" (Maldonado-Torres, 2008, p. 245), which in turn "provides the ground for the articulation of the *ego cogito* [I think]" (Maldonado-Torres, 2008, p. 245).

Thus, Biko was articulating a mode of being and becoming humanised in the context of centuries of racial subjugation. The becoming—obtaining dignity, power, authenticity, agency, relationality and all the qualities that embody humanness—is contingent on several factors: resistance to the negative ascriptions imposed by colonial discourses, practices and policies; insistence on an intellectual independence so that the hegemonic system does not define the rules of understanding reality and making knowledge; a critical engagement with African cultural concepts, spirituality, values and systems; and re-assertion of the history of African contributions to civilisation that simultaneously recognise the impositions and silences effected by colonialism.

Notwithstanding the possibilities that Black Consciousness offers for community psychology, these do—and necessarily must—remain in generative tension with Biko's revolutionary legacy signaling that discourses of humanism would be meaningless if they "were not part of a larger political philosophy and program of action" (Oliphant, 2008, p. 229). Herein precisely lies the dialectic of how Biko and Black Consciousness simultaneously resist (community) psychology's entanglement with coloniality. After all, a mere "psychologisation" of Black Consciousness but only dislocates it as a movement of change and denudes it of the very radical emancipatory principles to which we turn in this chapter. Biko's African humanism is ultimately forged on the *psycho-political* import of liberation from oppression, where praxis and activism are fundamental to the assertion of black humanity.

Conclusion

The archive has to be rich, varied and in a sense 'eclectic' enough to bear the weight of different contested interpretations and to allow them to battle out their differences in relation to the different texts and inter-texts which the archive itself makes available.
(Hall, 2001, p. 92)

This chapter contributes to the retrieval and recentering of the subjugated (community) psychologies of Africa as constituted by the Black Consciousness archive, and seeks to expand the liberatory potential of community psychologies founded on a decolonial attitude and situated knowledge standpoint. In Ngũgĩ wa Thiong'o's (1986) terms, this decolonising project of recentering is fundamentally about refuting the colonial archive as the derivation of Africa's consciousness(es). The Black Consciousness archive offers us profound and authoritative insights on the ever-urgent imperative of transformation in (community)psychology. Biko's culturally constitutive ontology, an element of which we conceptualise here as collectivist and humanising ontology, seeks to break from the colonial conception of being, and challenge its associated erasures and negations of black personhood. Our theorisations thus strive to transcend the extrusions of the colonial archive, an imperative that remains exigent in the context of the coloniality of knowledge, power and being. We recognise that ours is one account—that requires elaboration, refinement and revision—just as Black Consciousness is one articulation of the black experience. Notwithstanding, for the epistemo-political decolonisation of community psychology, "we must stake out our own discursive and political positions ... as African subjects of knowing" (Dei, 2012, p. 103).

We conclude in anticipation of the possibilities that Africa's knowledge archives may offer as we enlarge the ongoing and incomplete work of reconstructing community psychologies.

> The most important contribution on the black experience will have to come from Africa. For it is here, on the African continent, where the great and intricate drama of being-black-in-the-world is taking place. (Manganyi, 1973, p. 3)

References

Almeida, S. (2015). Race-based epistemologies: The role of race and dominance in knowledge production. *Wagadu: A Journal of Transnational Women's and Gender Studies, 13*, 79–105.

Appel, S. W. (1989). "Outstanding individuals do not arise from ancestrally poor stock". Racial science and the education of Black South Africans. *Journal of Negro Education, 58*, 544–546.

Azanian People's Organisation (AZAPO). (n.d.). *Black Consciousness*. Retrieved from http://azapo.org.za/about-azapo/black-consciousness/

Barnes, B., & Siswana, A. (2018). Psychology and decolonisation. Special issue. *South African Journal of Psychology, 48*(3), 297–399.

Biko, S. (2004). *I write what I like*. Picador Africa.

Bulhan, H. A. (1981). Psychological research in Africa: Genesis and function. *Race & Class, 23*, 13–42.

Bulhan, H. A. (1985). *Frantz Fanon and the psychology of oppression*. Plenum.
Bulhan, H. A. (2015). Stages of colonialism in Africa: From occupation of land to occupation of being. *Journal of Social and Political Psychology, 3*(1), 239–256.
Cabral, A. (1979). *Unity and struggle: Speeches and writings*. New York: Monthly Review Press.
Carnegie Commission Report. (1932). *The poor White problem in South Africa*. Pro Ecclesia Press.
Carolissen, R. L., & Duckett, P. S. (2018). Teaching toward decoloniality in community psychology and allied disciplines. Special issue. *American Journal of Community Psychology, 62*(3–4), 239–509.
Carton, B. (1990, May). *Civilizing "civilized labour." Poor Whites, poor roads and poor relief in Cradock, 1926-1934*. Paper presented at the South African Research Program, Spring Seminar, Yale University, New Haven, CT.
Cooper, S., Nicholas, L., Seedat, M., & Statman, J. (1990). Psychology and apartheid: Struggle for psychology in South Africa. In N. J. Nicholas & S. Cooper (Eds.), *Psychology and apartheid: Essays on the struggle and the mind in South Africa* (pp. 1–21). Johannesburg, South Africa: Vision/Madiba Press.
Cooper, S., & Ratele, K. (2018). The Black Consciousness psychology of Steve Biko. In S. Fernando & R. Moodley (Eds.), *Global psychologies: Mental health and the Global South* (pp. 245–260). Palgrave Macmillan.
de Sousa Santos, B. (2018). *The end of the cognitive empire: The coming of age of epistemologies of the South*. Duke University Press.
Dei, G. J. S. (2000). Rethinking the role of Indigenous knowledges in the academy. *International Journal of Inclusive Education, 4*(2), 111–132.
Dei, G. J. S. (2012). Indigenous anti-colonial knowledge as 'heritage knowledge' for promoting Black/African education in diasporic contexts. *Decolonization: Indigeneity, Education & Society, 1*(1), 102–119.
Dei, G. J. S. (2018). "Black like me": Reframing blackness for decolonial politics. *Educational Studies, 54*(2), 117–142.
Dineo Gqola, P. (2001). Contradictory locations: Blackwomen and the discourse of the Black Consciousness Movement (BCM) in South Africa. *Meridians: Feminism, Race, Transnationalism., 2*(1), 130–152.
Fanon, F. (1963). *The wretched of the earth*. Grove Weidenfeld.
Fanon, F. (1965). *A dying colonialism*. Grove Press.
Fanon, F. (1967). *Black skins, white masks*. Grove Press.
Goolam, N. M. I. (2006). The Timbuktu manuscripts - Rediscovering a written source of African law in the era of the African renaissance. *Fundamina: A Journal of Legal History, 12*(2), 29–50.
Gordon, L. R. (2000). *Existentia Africana: Understanding Africana existential thought*. Routledge.
Grosfoguel, R. (2013). The structure of knowledge in westernized universities: Epistemic racism/sexism and the four genocides/epistemicides of the long 16th century. *Human Architecture: Journal of the Sociology of Self-Knowledge, 11*(1), 73–90.
Hadfield, L. A. (2017). *Steve Biko and the Black Consciousness Movement*. Retrieved from http://oxfordre.com/africanhistory/view/10.1093/acrefore/9780190277734.001.0001/acrefore9780190277734-e-83
Hall, S. (2001). Constituting an archive. *Third Text, 15*(54), 89–92.
Hartman, S. (2007). *Lose your mother: A journey along the Atlantic slave route*. Farrar, Straus and Giroux.
Hook, D. (2004). Frantz Fanon, Steve Biko, 'psychopolitics' and critical psychology'. In D. Hook (Ed.), *Critical psychology* (pp. 84–114). Juta.
Letlaka-Rennert, K. (1990). Soweto street children. Implications of family disintegration for South African psychologists. In L. J. Nicholas & S. Cooper (Eds.), *Psychology and apartheid. Essays on the struggle for psychology and the mind in South Africa* (pp. 100–114). Vision.
Louw, J. (1986a). White poverty and psychology in South Africa: The poor White investigation of the Carnegie Commission. *Psychology in Society, 6*, 47–62.

Louw, J. (1986b). *This is thy work: A contextual history of applied psychology and labour in South Africa*. Unpublished doctoral dissertation. Amsterdam University, Amsterdam, The Netherlands.

Macas, L. A. (2004). Diversity and plurinationality. *Boletín ICC I-ARY Rimay, 6*(64). Retrieved from http://icci.nativeweb.org/boletin/64/macas.html

MacGaffey, W. (1981). African ideology and belief: A survey. *African Studies Review, 24*(2/3), 227–274.

Magaziner, D. R. (2011). Pieces of a (wo)man: Feminism, gender and adulthood in Black Consciousness, 1968–1977. *Journal of Southern African Studies, 37*(1), 45–61.

Maldonado-Torres, N. (2007). On the coloniality of being: Contributions to the development of a concept. *Cultural Studies, 21*(2–3), 240–270.

Maldonado-Torres, N. (2008). *Against war. Views from the underside of modernity*. Duke University Press.

Maldonado-Torres, N. (2016). *Outline of ten theses on coloniality and decoloniality*. Retrieved from http://fondation-frantzfanon.com/outline-of-ten-theses-on-coloniality-and-decoloniality/

Maldonado-Torres, N. (2017). On the coloniality of human rights. *Revista Crítica de Ciências Sociais, 114*, 117–136.

Maldonado-Torres, N. (2018). The decolonial turn. In J. Poblete (Ed.), *New approaches to Latin American studies: Culture and power* (pp. 111–127). Routledge.

Mamdani, M. (2012). *Define and rule: Native as political identity*. Harvard University Press.

Manganyi, N. C. (1973). *Being-black-in-the-world*. Ravan Press.

Mangena, M. J. O. (2008). The Black Consciousness philosophy and the women's question in South Africa: 1970-1980. In A. Mngxitama, A. Alexander, & N. C. Gibson (Eds.), *Biko lives* (pp. 253–266). Palgrave Macmillan.

Mazrui, A. A. (2005). The re-invention of Africa: Edward Said, V.Y. Mudimbe and beyond. *Research in African Literatures, 36*(3), 68–82.

Mignolo, W., & Walsh, C. (2018). *On decoloniality: Concepts, analytics, praxis*. Duke University Press.

Mngxitama, A., Alexander, A., & Gibson, N. C. (Eds.). (2008). *Biko lives! Contesting the legacies of Steve Biko*. Palgrave Macmillan.

Moodley, S. (Ed.). (2018). *Time to remember: Reflections of women from the Black Consciousness Movement*. Women for Awareness.

More, M. P. (2008). Biko: Africana existentialist philosopher. In A. Mngxitama, A. Alexander, & N. C. Gibson (Eds.), *Biko lives! Contesting the legacies of Steve Biko* (pp. 45–68). Palgrave Macmillan.

More, M. P. (2014). The intellectual foundations of the Black Consciousness Movement. In P. Vale, L. Hamilton, & E. Prinsloo (Eds.), *Intellectual traditions in South Africa: Ideas, individuals and institutions* (pp. 173–196). UKZN Press.

Mudimbe, V. Y. (1988). *The invention of Africa: Gnosis, philosophy, and the order of knowledge*. Indiana University Press.

Ndlovu-Gatsheni, S. J. (2013). *Empire, global coloniality and African subjectivity*. Berghahn Books.

Ndlovu-Gatsheni, S. J. (2018). *Epistemic freedom in Africa: Deprovincialization and decolonization*. Routledge.

Ngũgĩ wa Thiong'o. (1986). *Decolonising the mind: The politics of language in African literature*. J. Currey.

Nicholas, L. J. (Ed.). (1991). *Psychology and oppression: Critiques and protocols*. Skotaville.

Nicholas, L. J., & Cooper, S. (Eds.). (1990). *Psychology and apartheid*. Vision.

Oliphant, A. (2008). A human face: Biko's conceptions of African culture and humanism. In A. Mngxitama, A. Alexander, & N. C. Gibson (Eds.), *Biko lives* (pp. 213–232). Palgrave Macmillan.

Rodney, W. (1972). *How Europe underdeveloped Africa*. Howard University Press.

Seedat, M. (1993). *Topics, trends and silences in South African psychology 1948-1988: Ethnocentrism, crisis and liberatory echoes*. Unpublished doctoral dissertation, University of the Western Cape, Cape Town, South Africa.
Seedat, M. (1998). A characterization of South African psychology (1948-1988): The impact of exclusionary ideology. *South African Journal of Psychology, 28*(2), 74–84.
Seedat, M., & Lazarus, S. (2014). Community psychology in South Africa: Origins, developments, and manifestations. *South African Journal of Psychology, 44*(3), 267–281.
Seedat, M., & Mackenzie, S. (2008). The triangulated development of South African psychology: Race, scientific racism and professionalisation. In C. van Ommen & D. Painter (Eds.), *Interiors: A history of psychology in South Africa* (pp. 63–69). University of South Africa Press.
Seedat, M., & Suffla, S. (2017). Community psychology and its (dis)contents, archival legacies and decolonisation. Special Issue: Liberatory and critical voices in decolonizing community psychology. *South African Journal of Psychology, 47*(4), 421–431.
Senghor, L. S. (1994). Negritude: A humanism of the twentieth century. In L. Chrisman & P. Williams (Eds.), *Colonial discourse and post-colonial theory* (pp. 27–33). Columbia University Press.
Shariati, A. (1974). *Awaiting the saviour*. Free Islamic Literatures.
Tuğçe, K., Gómez, L., Molina, L. E., Dobles, I., & Adams, G. (2015). Decolonizing psychological science: Special Thematic Section on Decolonizing Psychological Science. *Journal of Social and Political Psychology, 3*, 213–413.
Vale, P., Hamilton, L., & Prinsloo, E. (Eds.). (2014). *Intellectual traditions in South Africa: Ideas, individuals and institutions*. UKZN Press.
van Ommen, C., & Painter, D. (Eds.). (2008). *Interiors: A history of psychology in South Africa*. University of South Africa Press.
Veriava, A., & Naidoo, P. (2008). Remembering Biko for the here and now. In A. Mngxitama, A. Alexander, & N. C. Gibson (Eds.), *Biko lives* (pp. 233–251). Palgrave Macmillan.
Wilderson, F. B. (2008). Biko and the problematic of presence. In A. Mngxitama, A. Alexander, & N. C. Gibson (Eds.), *Biko lives* (pp. 95–114). Palgrave Macmillan.
Yen, J. (2007). A history of "community" and community psychology in South Africa. In N. Duncan, B. Bowman, A. Naidoo, J. Pillay, & V. Ross (Eds.), *Community psychology: Analysis, context and action* (pp. 51–66). UCT Press.
Yen, J. (2008). A history of community psychology in South Africa. In C. van Ommen & D. Painter (Eds.), *Interiors: A history of psychology in South Africa* (pp. 385–412). University of South Africa Press.
Zeleza, P. T. (2007). The pasts and futures of African history: A generational inventory. *African Historical Review, 39*, 11–24.

An Orienting Conversation on Africa(n)-Centred Decolonial Community Psychologies

Kopano Ratele, Garth Stevens, and Nick Malherbe

In this chapter we stage an orienting conversation whose ultimate aim is to focus on that *turn* on the intellectual and political road toward decolonial Africa(n)-centred community psychologies. Our approach here is twofold: first, in using conversation as a kind of collective knowledge-making device, we look to disrupt and stretch some of the ways by which community psychologies are written, practiced and thought about. Second, we draw upon this conversational method to imagine a decolonising, Africa(n)-centred community psychology through a critical, non-psychological knowledge canon. More specifically, we consider and dialogue over what the works of Edward Said, Walter Rodney and the Combahee River Collective are able to offer those who are looking to reimagine community psychologies through Africa(n)-centredness and decoloniality. In short, we ruminate on the potentialities of form (i.e. conversation) and substance (i.e. the non-psychological canon) for a decolonising Africa(n)-centred community psychology.

Our three-way conversation—which, of necessity, includes points of disagreement, objections, consensus, and concession—was not a face-to-face conversation over dinner, but an ongoing *series* of historical and current conversations. We were therefore challenged in this chapter to present our conversation in as clear a manner as possible while, at the same time, retain its fundamental spirit and liveliness. Accordingly, the chapter does not seek to represent the fullness of our conversation.

K. Ratele (✉) · N. Malherbe
Institute for Social and Health Sciences, University of South Africa, Cape Town, South Africa

South African Medical Research Council-University of South Africa Masculinity and Health Research Unit, Cape Town, South Africa
e-mail: kopano.ratele@mrc.ac.za

G. Stevens
Department of Psychology, University of Witwatersrand, Johannesburg, South Africa

G. Stevens, C. C. Sonn (eds.), *Decoloniality and Epistemic Justice in Contemporary Community Psychology*, Community Psychology,
https://doi.org/10.1007/978-3-030-72220-3_3

Attempting to do so would, we felt, make for a cumbersome, tedious read. Rather, the chapter, organised around a set of questions, is a refined product of our conversation. We do not answer the various questions in any particular way, and sought instead to capture how the flow of ideas cohered and collided with one another. The questions, themselves, closely relate to a number of key terms in this book, including decoloniality, knowledge production, epistemic justice, interdisciplinarity, and community psychology. Each question is used to trigger reflections which lead to new or reworked insights, which may then lead to potential action. As is perhaps expected, when condensing a three-way conversation into a single prose, a number of topics are covered quite rapidly. We might say, then, that this moving conversational method resembles Michael Hardt and Antonio Negri's (2004) characterisation of the multitude as *singularities acting in common*, struggling toward the democratic ideal. Such non-linear conversational movement is, we will argue, a useful method in thinking Africa(n)-centredness and decoloniality into community psychologies.

A key objective of our conversation was to disrupt epistemic hegemony within community psychologies and to work toward epistemic justice in the very process of dialogically constructing arguments and knowledge-making through this iterative process. At the same time, the conversation is used to advance theses, observations and positions regarding epistemic experiments toward cultivating and claiming a space for Africa(n)-centred decolonial community psychologies within a field that continues to be dominated by Euro-American epistemes. What follows in this chapter is a frozen temporality of sorts, then, and encompasses some of the salient points to emerge in our conversations at this time—points of relevance to expanding the field of community psychology; engaging with what it means to be Africa(n)-centred in this pursuit; and drawing on decolonial understandings of the reticulations of power in a world that continues to be burdened by the legacies of coloniality.

As *singularities acting in common*, each of us have very different histories and positionalities, all of which informed how we engaged and reflected on the questions which organise our conversation in this chapter. What was important here were our respective loci of experience and enunciation, that is, how each speaker's conversational contribution was poised at the intersection of epistemology and a politics of location (Mignolo, 1999). While we do not explicitly draw on our different positionalities in our conversation, these positionalities undoubtedly shaped the conversation. Thus, because the histories of our voices and biographies were so crucial in how we used conversation as a knowledge-making device (see Bourdieu, 2000, for a well-known example of how conversation has been used in the context of scholarly inquiry), they are worth articulating, albeit somewhat briefly. Kopano is what the legal history of colonial and apartheid South Africa has variously referred to as native, non-European, non-white, Bantu, Black, or African; and in post-apartheid what is termed as African, Black and, a term he revolts against, Black African. Linguistically, he grew up with Sesotho as mother-tongue, although his father has roots in the land of the Zulus. His early life was in a small village on the north-west part of the country, a place that was incorporated under Bophuthatswana—literally, the land that gathers the Tswana—under spatial apartheid's Bantustan policy, and therefore Setswana was his first medium of instruction. In his pre-teens till his early twenties he lived in the East Rand in a township around 30 kilometres outside of Johannesburg that saw incredible violence in the late 1980s and early 1990s. Along with violence, his thought is informed by ongoing engagement with (dis)possession, (dis)location, (dis)connection in dialogue with struggles with and for power, being, and freedom. Seen as defining particulars of Black lives in South Africa these problems are for him sources to grapple with in his work as a critical/cultural psychologist with a major interest in the violence of and against men. Garth, in most parts of the world, would be considered to be of mixed-race descent, but self-identifies as Black as a consequence of his exposure to the political influences of Black Consciousness in the 1980s. His life experiences equally straddle both apartheid and post-apartheid South Africa, informing a deeply historical lens on contemporary apprehensions of the world. Trained initially as a clinical psychologist, his enduring interests have nevertheless been in critical race studies and in critical violence studies, both of which were, and remain, central to comprehending South Africa's contemporary axes and cleavages of difference, marginalisation, inequality and exploitation. As such, his approach has always understood psychology as a site and object of critique, disruption and evolution, away from complicity with marginality and in the service of the social good; but also as a vehicle for potential social relevance and transformation toward a more equitable global social system. Nick identifies (and indeed is identified) as a white middle-class male, and is thus imbued with a particular set of material and symbolic privileges and licenses that cohere with the global history of coloniality, as well as the particular manifestation of this history in South Africa. Although trained in community psychology, he oftentimes feels ambivalent about the discipline and the ways by which most its mainstream variants (and even many of its so-called critical approaches) act to bolster the dehumanising and exploitative logics of capitalism and coloniality. His psychologically-oriented work focuses largely on how to bring community psychology into the orbit of progressive politics, how to will community psychology to the demands of such a politics, and how to prevent community psychology from co-opting and deflating these politics.

The questions that we ask in this chapter ultimately go beyond the terms enunciated in the title of the volume and extend to other important associated terms, including but not limited to, poverty, feminism, the university, representation, racism, inequalities, violence, Africa, African psychology, class, and gender. But even within this framing, the specific questions used to direct our reflections and observations were not simply selected in an arbitrary or *laissez-faire* manner. Instead, they emerged from our individual and collective experiences of questions that have and continue to be posed within our discipline in relation to community psychology, decoloniality and our lives and studies as people of Africa. The questions therefore not only represent pivot points in our encounter amongst ourselves, but also historical encounters with real interlocuters, and future encounters with imagined publics for whom these questions continue to be important. The questions illuminate alternative methods as a site for epistemic justice; address coloniality and its persistence within contemporary systems of social organisation; what we mean by

Africa(n)-centred; why a decolonial approach offers a promise of liberatory praxis in this moment and how it can useful in addressing the recalcitrance of coloniality; and how such an approach relates to an existing dominant canon in psychology today. Thus, the chapter not only offers our observations, reflections and theses around community psychology's decolonial turn, but also reaches for new decolonial modes of engagement that exist beyond readily available and established disciplinary-bound processes that are considered to be valid forms of knowledge production.

What Kind of Knowledge Device is Conversation, and is it Really a Meaningful Way of Creating Knowledge?

Conversation is a legitimate route of making knowledge, a method that can lead to understandings and, in the to-and-fro of asking, can generate further dialogue, listening, and insights (see, e.g., Bourdieu, 2000). Often, it is the most meaningful of means for developing understanding. It is both suitable and valid as a mode of enquiry and knowledge-making, not just for community-centric psychologists, not just for a book such as this, but for all researchers (see Feldman, 1999; Seidman, 1991). Conversation is central to interviewing, but in contrast to the interview, a conversation does not predetermine the roles of the different speakers (i.e. "interviewer" and "interviewee"), with all interlocutors able to ask questions and offer answers. Conversation, in this regard, resembles the *inter-view*, wherein there is a shared awareness of dual viewing and contribution (Raz, 2005).

The idea that a conversation is a dynamic site for constructing knowledge is certainly not a new one. Habermas (1984), and other critical theorists associated with the Frankfurt School, insisted that in dialogue, communicative action allows for a consensus of sorts to develop around validity claims about the world, thus becoming the basis for social action. Freire (2005/1970) argued for the use of dialogic action to support new forms of (un)learning, understanding and freedom in a world that has been concealed by the workings of ideology. And, of course, Bakhtin (1981) made an argument for an infinite set of iterations that come from dialogue—as any dialogue is always premised on a previous dialogue, thereby generating a chain of new understandings which are evolving and changing in the context of the new dialogic production.

In the context of this text on community psychology, our dialogue on conversation raised a vital epistemological and methodological question, namely, whether conversation is a suitable, valid and robust method for constructing alternative, decolonial, Africa(n)-centred community psychologies? The simple answer is, it is. Apprehended as a method of enquiry and knowledge creation, conversation is well-suited for the task precisely because it is able to disrupt the hegemonic Cartesian forms of knowledge; decolonise the image of the solitary thinking man as a repository of ideas and creativity; surface collaborative enquiry and knowledge production; provide an opportunity for self- and other-reflection; and demystify how knowledge is made. Conversation, in its spoken form, is also obviously accessible to orality, to oral cultures, to non-academic settings, and to groups and individuals whose literacy levels are low, or who simply prefer the oral form (see for example the Latin American work on *testimonio* by Delgado Bernal, Burciaga, & Flores Carmona, 2012; as well as the work on *storytelling* by Sonn, Stevens, & Duncan, 2013). For Ngũgĩ wa Thiong'o (2012), conversation is a site of transmission and re-interpretation of knowledge, and is not merely the raw material for the formal writing enterprise that is so revered and, ultimately, elevated in Western modernist forms of knowledge production.

We understand our particular conversation as a patchwork of different questions, dialogical snippets, ideas, diatribes and reflections, all of which speak to and depart from one another as well as the central issue at hand (i.e. decolonising Africa(n)-centred community psychologies) in particular ways. We therefore work, and converse, in the tradition of relational dialogue (e.g., Moraga & Anzaldúa, 1984) which approaches the knowledge-making enterprise through an interactive, always contested, mode of "building of mental structures, containers of memory, feeling, and desire" (Pizer, 2003, p. 191). However, we wish to stress once again that we offer in this chapter the thrust of our conversation, rather than the conversation as such, which is to say that translating our conversation into text required that the conversation take on a more coherent—but no less dynamic—form than it did in real-time.

As to how to analyse our inter-views, conversation is of course regarded as the basis of a range of qualitative analyses, including thematic, conversational, and discourse analyses. And yet, conversation need not always be subject to analysis, and can be engaged as knowledge in and of itself. We ought to keep a fruitful connection between conversation as knowledge-making device and the development of methods of interpreting conversation, such as methods that embody collaboration between researchers and other conversation partners or interlocutors, where the researcher is part of that conversation.

What is "The Thing"[1] to Which we Wish to Turn in our Conversation?

Conversation is always moving, turning this or that way, around or through some topics, away from one thing and toward another. What is "the thing" from which we wish to turn in taking a decolonial turn? That is to say, in what ways is our

[1] Given the centrality of the idea of coloniality in this chapter, when we touched on the 'thing', what came to mind for one of us was Aimé Césaire's (1972) *Discourse on Colonialism* and his argument that the colonised are "thingified", objectified and commodified in the process of dehumanisation and infra-humanisation. It was, however, thought that the thing in this question can be thought of as a deliberate stratagem of inversion of a "thingifying" process, to turn it against coloniality itself.

conversation in this (or another conversation, or indeed any other act, lesson, theory, therapeutic modality, method, finding, article, chapter, or book) contributing to the possibilities of emancipatory futures within community psychologies? "The thing" is coloniality—the object onto which we cast our gaze and from which we would like to turn away from. Coloniality, specifically the coloniality of power (which has to be spoken of in relation to other instantiations of coloniality, such as the coloniality of being, knowledge and gender), is, writes Quijano (2000, p. 218),

> that specific basic element of the new pattern of world power that was based on the idea of 'race' and in the 'racial' social classification of world population—expressed in the 'racial' distribution of work, in the imposition of new 'racial' geocultural identities, in the concentration of the control of productive resources and capital, as social relations, including salary, as a privilege of 'Whiteness'.

Coloniality is the underbelly of Western modernity and its associated forms of colonisation—the first truly global and transnational process in which uneven distributional regimes are premised on the hierarchisation of peoples as human, subhuman and non-human. But it is also a legacy that continues to propagate and morph in the contemporary world and in current systems of inequality. Coloniality inheres in Euro-American-centralising, racialised, gendered, economic world power but, equally important, it is the quintessence of the very categories of thought, in work, in social relations, and in identity-making. It is in this sense that coloniality institutes a regime of epistemic degradation that dehumanises, "thingifies" and *Others* particular peoples for the purpose of ensuring that their material life coheres with the ideals of racial, patriarchal, and capitalist Western modernity (Maldonado-Torres, 2016). Coloniality seeks to remove colonised subjects from history, while simultaneously relying on their material, social, cultural and reproductive labour as the engine of history (Rodney, 1972). Inclusive within coloniality's epistemic project is hegemonic psychology which, in its ever-spreading tentacles, branches, interests, and proliferation, maps well onto coloniality's drive to individualise the psychosocial subject while making obscure any considerations of relational interconnectedness to the world (Tuck & Yang, 2012).

Several turns and precursors to what comes to be referred to as decolonisation or decoloniality—from Africa, Asia, Latin America, Aotearoa and the Caribbean (Stevens, 2017)—are evident in the intellectual and political junctures and antecedents which, as a whole, shape the emancipatory project of anti-colonial and decolonisation struggles, which then get extended through decolonial thinking as a process and social activity. Decoloniality—that which we turn to—is an *otherwise* turn in answer to what the Puerto Rican philosopher Nelson Maldonado-Torres calls "the colonizing turn" (Maldonado-Torres, 2011, p. 1) in psychology and broader society. To be more precise, a decolonial turn—such as we are taking—is a key response to coloniality. Decolonial praxis extends beyond didactic relationality by seeking to offer visions of a decolonised world that is antithetical to all of coloniality, and thus must comprise an intense, collective and sustained creative effort toward actualising the ideal of egalitarian and emancipatory futures—materially, relationally and metaphysically.

When Walter Mignolo says (2016, p. vii) "decoloniality is a path to heal the wounds of coloniality", what we understand with respect to psychology today is that, because of the influence of coloniality, much of psychology has been complicit and has visited psychic wounds on Africans and Black peoples; has worsened the wounds inflicted by coloniality; and/or has done little to heal these wounds. In regard to community psychologies, decoloniality therein is a response to the naturalised intellectual domination of psychology by Euro-American thought/power—the proper name of that disciplinary domination being Western or Euro-American-centric psychology over Africa and Africans (Ratele, 2019). Here, a decolonial psychology must, of necessity, be relational and communitarian in its orientation, as well as deliberate in how it engages with attempts to build more just and egalitarian societies. It must also speak of experiences from a specific context and of such a context, but always in relation to experiences of and from other contexts. Furthermore, it must encourage the acknowledgement of these standpoints as valid—even if they are contested—while rejecting and disrupting knowledge hegemonies and the hierarchisation of peoples that may emerge as consequential to such knowledges. There are numerous and varied pathways through which to activate such a decolonial psychology, as evidenced in the recent scholarship dedicated to decolonisation and psychology (e.g., Barnes & Siswana, 2018; Boonzaier & van Niekerk, 2019; Carolissen & Duckett, 2018; Ciofalo, 2019; Kurtiş, Gómez, Molina, Dobles, & Adams, 2015; Macleod, Bhatia, & Liu, 2020; Ratele, 2019; Seedat & Suffla, 2017).

Why is there Still Epistemological Domination Decades After the So-Called Classic Era of Colonisation?

The former colonies continue to be, among several ills and failures, politically dominated, economically controlled, and linguistically alienated. Political domination and economic control are, however, unsustainable without cultural supremacy and epistemological dislocation. It is clear, then, that political independence has failed to bring an end to coloniality. The lingering supremacy of the colonial order is visible within new modes of capitalist hegemony, precisely when we realise that economic interests converge with political, cultural and epistemological interests, resulting in the fact that the oppressor can be physically absent, can even be dead, yet the oppressed continue to act as if that oppressor is still around. This is precisely how the interests of the United States of America (USA) and wealthy Western European nations continue to dominate African countries decades after political independence from colonisation. Some of these interests have a long history, beginning with the imperialist and colonial campaigns into Africa in the 1400s. Others have more contemporary origins, such as the dictates of neoliberal market ideologies within an era of fiscal globalisation and the global hegemony of Euro-American cultures. Sometimes overlapping, at other times opposed to each other, to put it inelegantly, the economic interests of the USA and Western Europe in Africa are in the interests of the USA and Western Europe. To put this in an ironic way: the West is not interested in Africa. This is to say, colonial powers lack interest in an Africa that is for the benefit of Africans.

It is common cause that decades after the heyday of national independence in African countries in the 1960s, many aspects of people's economic livelihoods, cultural existence and knowledges from universities and research centres continue to be neo-colonised (that is, structured by new forms of oblique, uncoupled and remote forms of colonisation). Indeed, neo-colonialists—or what Paulo Freire (2005/1970), the critical Brazilian educationalist, called "sub-oppressors"—are those individuals from the ranks of the oppressed whose "ideal is to be men [sic]; but for them, to be men [sic] is to be oppressors" (p. 45). The neocolonialist model of humanity is thus constructed by and for the oppressor, by and for the colonizer, and in this regard understands 'normal humanness' through the demands and prisms of the coloniality of power (Wynter, 2003).

The implication of this insight for decolonial community psychologies is that the legacy, reach and traction of coloniality even today, makes it possible for teachers to be complicit in the perpetuation, subjugation or domination of Africans if they uncritically teach colonial or neo-colonial Euro-American psychology in their classes, or if researchers employ colonial or neo-colonial Euro-American psychological concepts, models and interpretations in their studies. It is possible for community psychologists to unconsciously—even if they are able to avoid overtly oppressive behaviour—reproduce colonial oppression in their work by adopting approaches that marginalise and alienate the voices, experiences and know-how of those living under the heel of coloniality with whom they may work. The South African-based Zimbabwean decolonial thinker, Sabelo Ndlovu-Gatsheni (2013, p. 5), has argued that,

> one of the strategies that have sustained the hegemony of the Euro-American-constructed world order is its ability to make African intellectuals and academics socially located in Africa and on the oppressed side to think and speak epistemically and linguistically like the Euro-American intellectuals and academics on the dominant side. This trap has made it very difficult for African intellectuals and academics to sustain a robust and critical perspective of Euro-American hegemonic knowledge and the asymmetrical power relations it enables.

Sub-oppression implies, at best, complicity with the oppression of one's people—however one defines the oppressed group that was once oppressed. At worst, it means active subjugation, including imprisoning and killing opponents and those considered to be dissident. Many psychologists from the ranks of the historically colonised can inflict cognitive and epistemological violence on our clients, students, and research participants. Many of us are not always fully conscientised about our complicity in reproducing models of psychology which are informed by coloniality.

To be fair, not all of psychology is simply a representation, product or consequence of *colonialised* ontology, epistemology and methodology. However, we remain complicit with coloniality when we do not speak out and act against the subtle (or, rather, veiled) violences within psychology (e.g. cultural and symbolic violence), as well as overt forms of physical, sexual and economic violence. In other words, along with rejecting ambitions to become sub-oppressors that capitulate to and adopt colonial representations of the colonised and incorporate the habits of the coloniser, or violent patriarch, or racist, or exploitative capitalist, it is the passive and effective support of and complicity with the power of the off-shore model of psychology that has to be overcome. Euro-American-centric psychology, which is the hegemonic model of psychology in many universities in Africa, was *not*—and in many cases *is* not—fundamentally intended to serve the emancipatory interests of the majority of people on the continent. It may still be the prevailing case that in many university psychology departments in Africa, and in numerous psychological studies conducted in African countries or about African peoples, Euro-American psychology—as opposed to decolonial Africa(n)-centred psychologies—forms, invites or induces people into positions of oppressors in the image of the former colonisers.

What Can Decolonial Africa(n)-Centring Community Psychologies Do to Resist Dominant Euro-American Models of Community Psychology?

Embedded in this question is an inquiry into what the fundamental touchstones of decolonial Africa(n)-centred community psychologies are. While references to what we mean when we talk about decoloniality are diffused throughout this chapter and book already, and a delineation of what constitutes community psychology abound both in South Africa and across the world (Reich, Riemer, Prilleltensky, & Montero, 2007; Stevens, 2007), it is the idea of being Africa(n)-centred that is perhaps most salient to clarify here.

Euro-American-centric community psychology was and is an off-shore body of knowledge (Mbembe, 2015). Such knowledge oftentimes coheres with epistemic violence, which occurs when dominating groups seek to fix representations of the *Other* in a manner that can be moulded to and for colonial purposes (Said, 1978). Epistemic violence in community psychology occurs when Euro-American ideas and practices seek to dominate thinking about communities and community psychologies outside of Euro-American contexts. Epistemic violence persists even when other, equally or more viable, interpretations are available (Teo, 2010), thereby signalling the symbolic valences that are attached to certain knowledge systems, and how these are undergirded by forms of social, political, cultural, ideological and economic capital.

The dominance of an off-shore, colonialised, and epistemologically violent model of knowledge is the principal reason that we have little concerning Africa in dominant models of (community) psychology, although we have psychology professors at universities in Africa (including professors from the ranks of the historically oppressed), large classes in psychology, thriving therapy practices, as well as a clutch of psychology conferences and seminars. Similarly, while there is considerable psychological research that comes out of Africa, many psychology journals and periodicals work to 'discipline the discipline', ensuring that only that which coheres with dominant, Euro-American knowledge forms is admitted into, and legitimised within, psychology's epistemic parameters.

The need, it seems, is to radically centre notions of Africa within community psychology, both internal to its epistemic core and within the possibilities of its praxes. Here, Africa(n)-centred is not about geographical location, but about epistemic situatedness—capturing the experiential, phenomenological and material fields that are of and from Africa, but speaking constantly to adjacent fields outside of Africa. To be clear, it is different from, although related to, what is commonly constituted as Black Psychology or Indigenous Psychology or African Psychology—psychologies that speak to the Black experience, the intersection of psychology with indigenous knowledges, and a psychology that is interconnected with ideas of African philosophy and/or cosmology (Stevens, 2015).

Coloniality, neo-colonial governance and epistemic violence are, however, never total. There is always resistance to colonial power, much of which seeks to create liberatory futures within the very act of resisting (James, 2012). Such resistance often assumes modalities that are deemed violent by dominant social groups (see Mignolo, 2009). Radical voices and dissidents in African countries and other parts of the world that have endured colonialism have always stood up and spoken out—often at great cost—against old and contemporary slavo-centric, imperialist, colonialist and other exploitative systems instituted by the enslavers and oppressors. The histories of these resistance efforts remain largely undocumented, or indeed marginalised within elitist or myopic historiographies (see James, 2012 and Rodney, 1972 for two outstanding exceptions here). There are people within psychology in Africa and around the globe who have spoken out against the dominant model of humanity that was birthed by the spokespersons of the old and new slave-masters, (neo)colonialists and apartheid fathers. Voices and spaces for dissent and resistance in psychology have always existed (e.g., Anonymous, 1986; Manganyi, 1973; Montero, 2002; Parker, 1989), but these spaces and voices have been small. Decolonial Africa(n)-centring community psychologists and community psychology students need to align themselves with other activist, decolonial and critical voices to create alternative and organic epistemic communities of practice and action, and to resist epistemological, cultural and material forms of dissonance and injustice.

While mainstream Euro-American-centric community psychology presents a history that in many ways aligns with colonial modernity (see Gokani & Walsh, 2017), there has also been a global tradition of anti-imperialist activist community psychologies (see Coimbra et al., 2012; Fryer, 2008; Montero, 1996; Seedat & Lazarus, 2011). Increasing spaces and supporting voices for a liberatory, decolonising psychology, and building upon this opposition against oppressive models of psychology, is what ought to be the concern of contemporary enactments of decolonial Africa(n)-centred community psychology. An important issue for community psychology today is the fact that the *model* of humanity—let us perhaps speak in the way psychology speaks and say, the model of the *individual*—arising from slavocracy, imperialism, colonialism and apartheid was intended to control how subjected people think, what they know, and what meanings they give to their experiences (see Wynter, 2003). In this model of humanity is an ethics of alterity, hierarchisation and exploitation in which violence replaces an ethics of relationality and reciprocity. The objective of slavocatric, imperial, colonial or apartheid psychology is not to enlighten the person and the community. It is not to open up the world. It is to miseducate, restrain, and manipulate those who receive such a psychology. It always keeps an eye to perpetuating global inequalities, be it on the basis of income, nation, gender, race, culture, sexuality, or religion. A psychology of this kind does not help people to explain their world or to change it, but reduces the politics of systemic colonial violence to a hermeneutic of individualised responsibility (see Parker, 2007). Here, too however, the control is not absolute; resistance is immanent. And within this resistance, a new model of humanity is central to how a decolonising psychology challenges the influence of the modernist conception of humanity that has at its core the differential categorisation of our species into humans, not-quite-humans and non-humans (Weheliye, 2014).

How Might we Create Space for Decolonial Africa(n)-Centredness in Community Psychologies?

Perhaps it is more suitable to ask how we can claim, as opposed to create, such a space within contemporary community psychologies. In the first instance, claiming a space for decolonial Africa(n)-centred community psychologies has to be based on a critical awareness and consciousness that community psychology itself is an internally variegated, contested, and sometimes, combative space—at times more aligned with mainstream psychology, and at other times being more critical of and, to varying degrees, rupturing with mainstream psychology. Differently composed and variously situated, community-oriented psychologies have been, and continue to be, practiced throughout the world and the African continent (Lazarus et al., 2006; Reich et al., 2007; Stevens, 2007). Some explicitly challenge systemic modes of oppression—as in South Africa and Egypt (Carrillo, Welsh, & Zaki, 2015; Seedat & Lazarus, 2011)—while others engage less overtly with political models, such as the modes of community psychology practiced in Kenya, Zimbabwe and Cameroon, where emphasis is placed on local knowledge production (Lazarus et al., 2006; Nsamenang, Fru, & Browne, 2007). Thus, the epistemological, cultural and political orientations of community psychologies have always been far from monolithic.

In the second instance, there is the need to mobilise, activate and build on those elements within community psychology that lend themselves to an Africa(n)-centred orientation. Indeed, mental health practices throughout Africa, including community psychology, have in different ways sought to accommodate the concept of 'community'. It is therefore within the tradition of community psychology (not always named as such) as it is practiced in Africa that we may begin to articulate a decolonial Africa(n)-centred community psychology which is sensitive to local knowledge systems and practices, while critically and constructively engaging Euro-American psychological theory in a manner that resists oppressive and institutional hegemony (Seedat, 1997). Claiming space for such a psychology requires us to problematise the static ways by which communities are often conceived, that is, as wholly virtuous or entirely dangerous (see Butchart & Seedat, 1990); patronised or pathologised. While we do not advocate the rejection of universal proclamations in every instance (some facets of community life surely deserve valorisation), it remains imperative for any decolonial psychology to take seriously the particular ways by which 'community' is discursively constituted within, by, and for the discipline (see Coimbra et al., 2012), and how such constitution influences the psychological, physical and material well-being of actually existing communities. In other words, various constructions of community contribute either to the reproduction of community as the relegated domain of the marginal, or the periphery, of a society, while others have the capacity to think about community as relational and interconnected sites of meaning-making, knowledge-production and social action which are located epistemically within particular contexts. It is the elevation of the latter conception that may be more productive in claiming space for Africa(n)-centred community psychologies.

In the third and final instance, claiming such space must be a deliberate and conscious political act. Knowledge is not an objective artefact that is simply to be uncovered or discovered, and it is always (re)produced and imbued with power. This is to say, the kinds of knowledges that become hegemonic are based in part on the power to create such knowledge, and that knowledge itself becomes a site for power to be wielded (Foucault, 2008; Said, 1978). But, as we have argued, power also generates resistance and, as such, inserting a different, alternative, and historically subordinated knowledge into such a space is an act of political and epistemic contestation that brings with it the possibility for cognitive and epistemic justice (de Sousa Santos, 2016). This is central to the praxes of decoloniality—to contest the matrix of power, knowledge and being that has been constructed within and through colonialism and its legacies.

Of what Relevance is the Canon to the Development of Decolonial Africa(n)-Centred Community Psychologies?

Chris Barker (2004, p. 19) speaks of the canon as "a body of work held to be the most important within a particular tradition, or sometimes to the parameters that surround the work associated with a given author". Although canons can be useful, their necessary exclusions and blind spots can also serve to falsely portray the totality of legitimate knowledge. Raewyn Connell (2019) notes that the sociological canon was consciously constructed in the twentieth century to "allow—indeed invite—argument within a shared space" (p. 355), and thus limit the scope, boundaries and language that is made available to revered sociological knowledge. While it would not be accurate to say that truly radical ideas always resist canonisation, such ideas are frequently canonised myopically. For instance, John Rawls is far better remembered for his liberal moral philosophy than his latter-day socialist ideals (Edmundson, 2017); just as W. E. B. Du Bois is little remembered for his anti-imperialism (Connell, 2019), socialism, and principled stance on gender equity (Davis, 1983). Further still, the radical politics of Martin Luther King Jr. (West, 2018) and Nelson Mandela (Ndlovu-Gatsheni, 2015) are repeatedly erased within official social remembering efforts. Canonisation oftentimes works to feed into liberal mythologies surrounding *Great Men of History* at the neglect of histories from below, that is to say, the multitude as the driving historical force (see Hardt & Negri, 2004).

Psychology, of course, presents its own (myopic) canon. This includes critical community psychologies, for which radical psychological thinkers and activists like Frantz Fanon (1963, 1967) and Ignacio Martín-Baró (1994) are often central. While there is certainly much to learn from thinkers like these, the expansive (but never complete or entirely omnipotent) system of coloniality requires that decolonial enactments of Africa(n)-centered community psychology look beyond canonisation, and particularly canonisation within psychology. Attempts to decolonise psychological praxis should not see us continually reproducing the discipline through the most readily available epistemes. We are instead urged to stretch the emancipatory capacities of community psychology through transdisciplinary engagements that may contort it in ways that we do not expect and that do not cohere with how community psychology has been canonically constructed. It is perhaps only in this way that we can begin to will psychology (including community psychology) toward the kind of radical politics to which it has so often been resistant.

To begin stretching, bending, contorting and even re-shaping community psychology in this way is challenging, and requires a form of creative labour on the part of psychologists working with communities. As we have already noted, we are seeing today an increasing body of scholarship that seeks to address and articulate issues of (de)coloniality within community psychology (see, e.g., Boonzaier & van Niekerk, 2019; Carolissen & Duckett, 2018; Seedat & Suffla, 2017). During our conversation on decolonial Africa(n)-centred community psychologies, we repeatedly drew on this work as well as particular thinkers and bodies of thought that have been meaningful to each of us. We did not select these thinkers in accordance to a specific set of criteria. Whether by accident of birth or peculiarity of time, the different figures came to influence our respective thinking (and, in the context of our conversation, collective thought) in different ways. Some of these figures have been more influential to some of us than they have been to others. There were three bodies of work in particular that we referenced in our conversations, namely that of the Combahee River Collective, Walter Rodney and Edward Said, precisely because they are not commonly cited in community psychology. Of particular interest, we wish to note that of these three bodies of work, only Rodney undertook work that was explicitly rooted in Africa. Nonetheless, we wish also to argue that this work is able to influence how we think of, envision and enact decolonial community psychologies within and for Africa in the context of the world, and should thus be thought with, against and through other Africa(n)-centred decolonial knowledges.

While the work of Edward Said, Walter Rodney and the Combahee River Collective should certainly not be considered the beginning or end points of thinking through decoloniality from within African-centred community psychologies, our engagement with them serves as a nascent attempt to begin to develop these psychologies from outside of the psychological canon. In what follows, we offer a somewhat synthesised presentation of the conversations that we had concerning the relevance of this work for decolonial Africa(n)-centred community psychologies.

Edward Said was a Palestinian scholar-activist who was born in Jerusalem in 1935 and died in 2003 from leukaemia. His impressive oeuvre had a profound influence on post-colonial studies, and spans a great number of topics, including literacy criticism, the social role of intellectuals, music, Palestinian justice, culture and humanism. Said is undoubtedly best known today for his beautifully-crafted and meticulously researched book, *Orientalism* (see Said, 1978), and to a lesser extent for its sequel, *Culture and Imperialism* (see Said, 1993).

In *Orientalism*, he argues that the field of area studies known as Orientalism (which, largely due to the influence of Said, is rarely named as such today) tends to depict the Middle East, Asia and North Africa in racist and patronising ways that facilitate colonial control. Put differently, the so-called Orient is discursively constituted by European scholarship in ways that seek to justify control of the Orient along elite categories and definitions. Said's nuanced analysis reveals that the Orient is not, in every instance, constructed as inferior to the Occident, but is always an essentially different Other. In a typically acerbic turn of phrase, Said notes that the Orientalist's "presence is enabled by the Orient's effective absence" (Said, 1978, p. 208). In this way, the (Oriental) Other ceases to exist without reference to the (Occidental) self.

Said's (1978) thesis in *Orientalism* is not exclusively confined to the so-called Orient, and holds relevance for all area studies, including those concerning Africa. Although engaging with the work of Freud (Said, 2003), we find no evidence that Said ever drew from community psychology. His mode of thinking through issues of representation is, however, instructive for those of us looking to locate a decolonising politics within Africa(n)-centred community psychologies. It is Said's (1978) humanistic insights that are so important here, particularly with respect to histories of Euro-American psychological thinking that delineate one psychology for the European, and another (often crafted through stereotype) for the African (Ratele, 2019). As is stressed in *Orientalism*, material politics that reject oppressive knowledge forms and ways of thinking begin in language, making how we represent ourselves and others (as well as ourselves to others, and others to ourselves) a crucial consideration for decoloniality. Projects of self-determination within Africa(n)-centred community psychologies should seek to assert forms of epistemic resistance (and correction) through new and fundamentally humanising discursive frames. In addition to learning from the form of *Orientalism* (exhaustively researched; rhetorically stunning; driven by conviction), the book's content is able to assist us in navigating the disempowering discursive landscape within which people are made to articulate their inner lifeworlds. Said's (1978) psychosocial, historical and epistemic framing points us toward the discursive conditions from which decolonising resistance politics can draw psychological strength.

Walter Rodney was a Pan-Africanist Marxist historian from Georgetown, Guyana, who was assassinated in 1980. A self-described "guerrilla intellectual", Rodney was a committed activist who wrote prolifically and was well-known for his potent, commanding and clearly-articulated public lectures (see Rodney, 1969). Today he is best remembered for his 1972 book, *How Europe Underdeveloped Africa*, in which he painstakingly analyses how five centuries of slavery, colonialism and international capitalism (the latter of which we would speak of today as neoliberalism) have systemically underdeveloped African countries for the benefit of European superpowers. These superpowers, Rodney argues, effectively own the means of production in Africa, often working with and through the national bourgeois (see Fanon, 1963). Monolithic images of 'Africa' are destabilised through Rodney's meticulous account of how different environmental, colonial and superstructural elements cohered in particular African countries to give shape to specific oppressive circumstances that fed into wider colonial networks of exploitation. Careful never to strip the colonised of agency, Rodney's book demonstrates to us how underdevelopment destroyed local markets, cultures and political institutions throughout the African continent, all while fleshing out how the racial and patriarchal nature of capitalism functions. The revolutionary impetus of *How Europe Underdeveloped Africa* is its assertion that we need to understand systems of oppression in order to overthrow them.

Community psychology in Africa, it would appear, has rarely, if it all, looked to Rodney's (1972) work. This is undoubtedly to our detriment. It seems self-evident that psychological functioning is in large part determined by socio-historic circumstances. Yet, as Rodney's text demonstrates, a cursory familiarity with these circumstances is insufficient if we are to comprehensively understand how history reaches into the psycho-material character of the world we find ourselves in today. The clarity of Rodney's (1972) work—which, as is evidenced in his "groundings" pedagogic model (see Rodney, 1969), is intended for large, non-specialised audiences—assists us in de-naturalising the wretched social conditions to which the majority of people are subject. There is development only because there is underdevelopment. What is particularly inspiring for Africa(n)-centred community psychologist is the activist impulse that undergirds Rodney's analysis. Bringing Rodney's Pan-Africanist historical materialism into our psychological engagements with communities opens up critical space within which to remake psychology as a functionary of decolonial praxis. Like Rodney, community psychologists should strive toward a sharply analytical, systems-focused approach that resists collapsing into fatalist narratives, and is always geared toward a radical kind of hope.

In reaction to the myopia of white bourgeois feminism, a group of Black lesbian feminists from Boston, USA, formed in 1974 what became known as the Combahee River Collective (Harris, 2001). During the next 6 years, the group involved itself in a number of activist efforts (e.g. campaigns for workers' rights and equal education, as well as protests against police brutality and gender-based violence). Today, the

Combahee River Collective is best known for *A Black Feminist Statement* (see Combahee River Collective, 2000), which was published in 1977 and penned by three founding members of the Collective, Barbara Smith, Beverly Smith and Demita Frazier (Haider, 2018). The Statement is concerned with fostering an intersectional consciousness that is alert to the multifaceted nature of oppressions, as well as building political coalitions which are sensitive to flows of oppressive powers that act within and against identity. Perceptively, it notes that "the major systems of oppression are interlocking, [with] the real class situation [being one] of persons who are not merely raceless, sexless workers" (Combahee River Collective, 2000, p. 264). Although somewhat resembling (yet predating) Kimberlé Crenshaw's (1991) highly influential notion of intersectionality, the Combahee River Collective's Statement was not rooted in the academy. It was crafted primarily as a document that would inform an expansively conceived resistance politics which, although embedded in the materiality of struggle, rejected economic determinism (Haider, 2018). *A Black Feminist Statement* offers a useful framework for articulating an emancipatory politics which reject the oppressive logic of socially constructed identity categories, while always remaining sensitive to the materiality of identity-based oppression. The legacy of the Combahee River Collective's tireless coalition building—as well as its fiercely feminist, anti-racist and anti-capitalist politics (Harris, 2001)—is today observed in a number of radical and organised political collectives, such as the Movement for Black Lives in the USA, and Abahlali baseMjondolo in South Africa.

The Combahee River Collective's (2000) *A Black Feminist Statement* is seldom acknowledged within and by community psychology. This omission (along with the discipline's ignoring of other radical statements and manifestos) perhaps points to the fact that most critical community psychology engagements tend to prioritise social analysis over resistance politics proper. This is oftentimes linked to institutional arrangements that bar community psychology from immersing itself in activities that may be deemed 'violent' by dominant social arrangements (see Gokani & Walsh, 2017; Mignolo, 2009). Africa(n)-centred community psychologists oriented toward decoloniality are thus presented with a deadlock: how are we able to fully embrace decoloniality from within a discipline that so often prevents us from doing so? The answer lies in how we navigate a strategic and an ever-shifting embrace of our professional identities. As psychologists, we can and should draw from the available disciplinary resources that can practically facilitate the conditions necessary for organising collective decolonial resistance (Ratele et al., 2018), while offering our solidary to these resistance efforts as citizens, rather than institutionally-embedded psychologists (Gokani & Walsh, 2017). In addition, then, to contributing to community-led change-making efforts, *A Black Feminist Statement* reminds community psychologists that such change-making must, in creative ways, be prioritised over institutional requirements. Those looking to construct decolonial Africa(n)-centred community psychologies should draw inspiration from, and strive to work with, radical organisations like the Combahee River Collective. Indeed, it is organisations like these that must guide the shape and form assumed by a decolonial African psychology that seeks to legitimise itself to communities, rather than to the institutional requirements of coloniality.

Conclusion

Our goal in this chapter (expressed through both its content and its form) was to compose an extended conversation on how Africa(n)-centred community psychologies are able to move with, rather than against, the so-called *decolonial turn*. Along the way we encountered the familiar kinds of challenges faced when attempting to synthesise and structure such an anarchic, critically-oriented, conversation. The question-and-answer and reflective style of this conversation, we feel, spoke to how we sought to grapple with what emerged as the key issues (e.g. poverty, the university, representation, racism, inequalities, violence, Africa itself, African psychology, class, and gender) facing Africa(n)-centred decolonial community psychologies.

Our conversational method is, of course, more provocative than it is analytically rigorous, and thus should be assessed on the basis of whether it did, indeed, provoke thinking around the decolonising, Africa(n)-centred possibilities and potentialities of community psychologies. There are undoubtedly gaps, lapses, biases, blind spots and contestable ideas that we have offered here, all of which were apparent in the conversation itself. There may also have been better, clearer, ways by which to present a conversation such as this. Certainly, in shoehorning our conversation into this chapter's relatively coherent form, we may have lost some its original, and perhaps idiosyncratic, nuances, interruptions and contradictions. Furthermore, with respect to our consideration of the Combahee River Collective, Walter Rodney and Edward Said, we undoubtedly did not afford sufficient attention to this work, and how it can inform community psychologies. We therefore urge other psychologists of community who are concerned with Africa(n)-centredness and/or decoloniality to not only collectively think through, engage, contest and build upon the ideas offered in this chapter, but also to gapple with how these ideas came about.

We have begun in this chapter to articulate a number of pathways through which to imagine not only decolonising Africa(n)-centred community psychologies, but also how we might begin to locate community psychology within the broader decolonial project. It is hoped that this conversation extends the network, reach and imaginative horizons by which a world-centred community psychology that is within, for, by, and of Africa is able to embrace decoloniality.

References

Anonymous. (1986). Some thoughts on a more relevant or indigenous counselling psychology in South Africa: Discovering the socio-political context of the oppressed. *Psychology in Society, 5*, 81–89.

Bakhtin, M. (1981). *The dialogic imagination: Four essays*. Austin, TX: University of Texas Press.

Barker, C. (2004). *The SAGE dictionary of cultural studies*. London: Sage.

Barnes, B., & Siswana, A. (Eds.). (2018). Psychology and decolonisation [Special issue]. South Africa Journal of Psychology, 48(3).

Boonzaier, F., & van Niekerk, T. (Eds.). (2019). *Decolonial feminist community psychology*. Cham, Switzerland: Springer.

Bourdieu, P. (2000). *The weight of the world: Social suffering in contemporary society*. Stanford, CA: Stanford University Press.
Butchart, A., & Seedat, M. (1990). Within and without: Images of community and implications for South African psychology. *Social Science & Medicine, 31*(10), 1093–1102.
Carolissen, R., & Duckett, P. (Eds.). (2018). Teaching toward decoloniality in community psychology and allied disciplines [Special issue]. *American Journal of Community Psychology, 62*(3–4), 239.
Carrillo, A. M., Welsh, E., & Zaki, N. (2015). The revolution, new Egypt, and psychological sense of community. *Journal of Community Psychology, 43*(1), 107–118.
Césaire, A. (1972). *Discourse on colonialism*. New York: Monthly Review Press.
Ciofalo, N. (Ed.). (2019). *Indigenous psychologies in an era of decolonization*. New York: Springer.
Coimbra, J. L., Duckett, P., Fryer, D., Makkawi, I., Menezes, I., Seedat, M., et al. (2012). Rethinking community psychology: Critical insights. *The Australian Community Psychologist, 24*(2), 135–142.
Combahee River Collective. (2000). The Combahee River collective statement. In B. Smith (Ed.), *Home girls: A feminist anthology* (pp. 264–274). New Brunswick, NY: Rutgers University Press.
Connell, R. (2019). Canons and colonies: The global trajectory of sociology. *Estudos Históricos (Rio de Janeiro), 32*(67), 349–367.
Crenshaw, K. (1991). Mapping the margins: Intersectionality, identity politics, and violence against women of color. *Stanford Law Review, 43*(6), 1241–1299.
Davis, A. Y. (1983). *Women, race, and class*. New York: Vintage.
de Sousa Santos, B. (2016). *Epistemologies of the south. Justice against epistemicide*. New York: Routledge.
Delgado Bernal, D., Burciaga, R., & Flores Carmona, J. (2012). Chicana/Latina testimonios: Mapping the methodological, pedagogical, and political. *Equity and Excellence in Education, 45*(3), 363–372.
Edmundson, W. A. (2017). *John Rawls: Reticent socialist*. Cambridge, UK: Cambridge University Press.
Fanon, F. (1963). *The wretched of the earth*. New York: Grove Press.
Fanon, F. (1967). *Black skin, white masks*. New York: Grove Press.
Feldman, A. (1999). The role of conversation in collaborative action research. *Educational Action Research, 7*(1), 125–147.
Foucault, M. (2008). *The history of sexuality*. London: Penguin.
Freire, P. (2005). *Pedagogy of the oppressed*. New York: Continuum. (Original work published, 1970).
Fryer, D. (2008). Some questions about "the history of community psychology". *Journal of Community Psychology, 36*(5), 572–586.
Gokani, R., & Walsh, R. T. (2017). On the historical and conceptual foundations of a community psychology of social transformation. *American Journal of Community Psychology, 59*(3–4), 284–294.
Habermas, J. (1984). *The theory of communicative action* (Vol. 1: Reason and the rationalization of society). Boston: Beacon Press.
Haider, A. (2018). *Mistaken identity: Race and class in the age of Trump*. London: Verso.
Hardt, M., & Negri, A. (2004). *Multitude*. London: Penguin Books.
Harris, D. (2001). From the Kennedy commission to the Combahee Collective: Black feminist organizing, 1960-80. In B. Collier-Thomas & V. P. Franklin (Eds.), *Sisters in the struggle: African American women in the Civil Rights-Black Power movement* (pp. 280–305). New York: New York University Press.
James, C. L. R. (2012). *A history of pan-African revolt*. Oakland, CA: PM Press.
Kurtiş, T., Gómez, L., Molina, L., Dobles, I., & Adams, G. (Eds.). (2015). Decolonizing psychological science. [Special issue]. *Journal of Social and Political Psychology, 3*(1), 213.
Lazarus, S., Bojuwoye, O., Chireshe, R., Myambo, K., Akotia, C., Mogaji, A., et al. (2006). Community psychology in Africa: Views from across the continent. *Journal of Psychology in Africa, 16*(2), 147–160.

Macleod, C., Bhatia, S., & Liu, W. (2020). Feminisms and decolonising psychology [special issue]. *Feminism & Psychology, 30*(3), 343.
Maldonado-Torres, N. (2011). Thinking through the decolonial turn: Post-continental interventions in theory, philosophy, and critique—An introduction. *Transmodernity: Journal of Peripheral Cultural Production of the Luso-Hispanic World, 1*(2), 1–15.
Maldonado-Torres, N. (2016). *Outline of ten theses on coloniality and decoloniality.* Fondation Frantz Fanon. Retrieved from http://frantzfanonfoundation-fondationfrantzfanon.com/IMG/pdf/maldonado-torres_outline_of_ten_theses-10.23.16_pdf
Manganyi, N. C. (1973). *Being-black-in-the-world.* Johannesburg, South Africa: Ravan Press.
Martín-Baró, I. (1994). *Writings for a liberation psychology.* Cambridge, MA: Harvard University Press.
Mbembe, A. (2015). *Decolonizing knowledge and the question of the archive.* Rhodes University. Retrieved from https://www.ru.ac.za/media/rhodesuniversity/content/equityampinstitutionalculture/documents/Achille_Mbembe_-_Decolonizing_Knowledge_and_the_Question_of_the_Archive-1.pdf
Mignolo, W. (2016). Foreword/Decolonial body-geo-politics at large. In S. Bakshi, S. Jivraj, & S. Posocco (Eds.), *Decolonizing sexualities: Transnational perspectives, critical interventions.* Oxford, UK: Counterpress.
Mignolo, W. D. (1999). I am where I think: Epistemology and the colonial difference. *Journal of Latin American Cultural Studies, 8*(2), 235–245.
Mignolo, W. D. (2009). Epistemic disobedience, independent thought and decolonial freedom. *Theory, Culture & Society, 26*(7–8), 1–23.
Montero, M. (1996). Parallel lives: Community psychology in Latin America and the United States. *American Journal of Community Psychology, 24*(5), 589–605.
Montero, M. (2002). On the construction of reality and truth. Towards an epistemology of community social psychology. *American Journal of Community Psychology, 30*(4), 571–584.
Moraga, C., & Anzaldúa, G. (Eds.). (1984). *This bridge called my back: Writings by radical women of color* (2nd ed.). New York: Kitchen Table Press.
Ndlovu-Gatsheni, S. J. (2013). *Coloniality of power in postcolonial Africa: Myths of decolonization.* Dakar, Senegal: CODESRIA.
Ndlovu-Gatsheni, S. J. (2015). The decolonial Mandela: Embodiment of peace, justice, and humanism. *Journal of Developing Societies, 31*, 305–332.
Ngũgĩ wa Thiong'o. (2012). *Globalectics: Theory and the politics of knowing.* New York: Columbia University Press.
Nsamenang, A. B., Fru, F. N., & Browne, M. A. (2007). The roots of community psychology in Cameroon. In S. M. Reich, M. Reimer, I. Prilleltensky, & M. Montero (Eds.), *International community psychology: History and theories* (pp. 392–406). New York: Springer.
Parker, I. (1989). Discourse and power. In J. Shotter & K. J. Gergen (Eds.), *Texts of identity* (pp. 56–69). London: Sage.
Parker, I. (2007). *Revolution in psychology: Alienation to emancipation.* London: Pluto Press.
Pizer, B. (2003). When the crunch is a (k)not: A crimp in relational dialogue. *Psychoanalytic Dialogues, 13*(2), 171–192.
Quijano, A. (2000). Coloniality of power and Eurocentrism in Latin America. *International Sociology, 15*(2), 215–232.
Ratele, K. (2019). *The world looks this from here: Thoughts on African psychology.* Johannesburg, South Africa: Wits University Press.
Ratele, K., Cornell, J., Dlamini, S., Helman, R., Malherbe, N., & Titi, N. (2018). Some basic questions about (a) decolonizing Africa (n)-centred psychology considered. *South Africa Journal of Psychology, 48*(3), 331–342.
Raz, A. (2005). A note on inter-viewing: Using symbolic interactionism for interview analysis. *Studies in Symbolic Interaction, 28*, 323–339.
Reich, S., Riemer, M., Prilleltensky, I., & Montero, M. (Eds.). (2007). *International community psychology: History and theories.* New York: Springer.

Rodney, W. (1969). *The groundings with my brothers*. London: Bogle-L'Ouverture.

Rodney, W. (1972). *How Europe underdeveloped Africa*. Washington, DC: Howard University.

Said, E. W. (1978). *Orientalism*. New York: Pantheon Books.

Said, E. W. (1993). *Culture and Imperialism*. New York: Knopf.

Said, E. W. (2003). *Freud and the non-European*. London: Verso.

Seedat, M. (1997). The quest for liberatory psychology. *South Africa Journal of Psychology, 27*(4), 261–270.

Seedat, M., & Lazarus, S. (2011). Community psychology in South Africa: Origins, developments, and manifestations. *Journal of Community Psychology, 39*(3), 241–257.

Seedat, M., & Suffla, S. (2017). Liberatory and critical voices in decolonising community psychology [special issue]. *South Africa Journal of Psychology, 47*(4), 419.

Seidman, I. E. (1991). *Interviewing as qualitative research: A guide for researchers in education and the social sciences*. New York: Teachers College Press.

Sonn, C. C., Stevens, G., & Duncan, N. (2013). Decolonisation, critical methodologies, and why stories matter. In G. Stevens, N. Duncan, & D. Hook (Eds.), *Race, memory and the apartheid archive: Towards a transformative psychosocial praxis* (pp. 295–314). London/Johannesburg: Palgrave Macmillan/Wits University Press.

Stevens, G. (2007). The international emergence and development of community psychology. In N. Duncan, B. Bowman, A. Naidoo, J. Pillay, & V. Roos (Eds.), *Community psychology: Analysis, context and action* (pp. 27–50). Cape Town, South Africa: UCT Press.

Stevens, G. (2015). Black psychology: Resistance, reclamation and redefinition. In I. Parker (Ed.), *Handbook of critical psychology* (pp. 182–190). London: Routledge.

Stevens, G. (2017). Critical psychosocial mnemonics as a decolonising participatory method: Towards refiguring and reclaiming the archive through memory, stories and narratives. In M. Seedat, S. Suffla, & D. J. Christie (Eds.), *Emancipatory and participatory methodologies in peace, critical, and community psychology* (pp. 179–199). New York: Springer.

Teo, T. (2010). What is epistemological violence in the empirical social sciences? *Social and Personality Psychology Compass, 4*(5), 295–303.

Tuck, E., & Yang, K. W. (2012). Decolonization is not a metaphor. *Decolonization: Indigeneity, Education and Society, 1*(1), 1–40.

Weheliye, A. G. (2014). *Habeas viscus*. Durham, NC: Duke University Press.

West, C. (2018, April 4). Martin Luther King Jr was a radical. We must not sterilize his legacy. *The Guardian*. Retrieved from https://www.theguardian.com/commentisfree/2018/apr/04/martin-luther-king-cornel-west-legacy

Wynter, S. (2003). Unsettling the coloniality of being/power/truth/freedom: Towards the human, after man, its overrepresentation – An argument. *CR: The New Centennial Review, 3*(3), 257–337.

Decolonizing Critical Knowledges Borne in the Borderlands: From "Morbid Symptoms" to Critical Solidarities

Michelle M. Fine

Critical Inquiry in Crisis

> The crisis consists precisely in the fact that the old is dying and the new cannot be born; in this interregnum a great variety of morbid symptoms appear.
>
> Gramsci (1971)

Almost 100 years after Antonio Gramsci scribbled these lines in his prison notebooks in Italy, we are drenched, again as always, in morbid symptoms. This essay appreciates and worries that community/critical psychology has brilliantly exposed the effects of colonialism, capitalism, white supremacy and state violence but has been far less attentive to the systematic and simultaneous accumulation of privilege, power and profit. That is, in solidarity I fear we have unwittingly reproduced a downstream gaze.

Resurrecting activist-scholar ancestors who have toiled in the "borderlands" to produce knowledge alongside social movements, this essay calls up our buried legacy of twentieth century research collectives aligned with justice struggles to document who pays but also who benefits from structural violence; theorizing how injustice moves under the skin, into bodies and communities of poverty and wealth and interrogating what Jessica Ruglis and I (2009) have called circuits of oppression, privilege, and resistance.

M. M. Fine (✉)
Critical Psychology and Urban Education, CUNY, New York, NY, USA
e-mail: Mfine@gc.cuny.edu

G. Stevens, C. C. Sonn (eds.), *Decoloniality and Epistemic Justice in Contemporary Community Psychology*, Community Psychology,
https://doi.org/10.1007/978-3-030-72220-3_4

June 2018

In the name of "national security," the U.S. government has committed to a strategic racialized policy of snatching children from migrating families seeking asylum at the border. More than 3000 children, 13 months to 17 years, have been forcibly removed from the arms of their parents who have crossed deserts, rivers and dangerous borders to reach the shores of Texas or California. Border guards—many with lineage from the same countries of origin—are locking children in cages and deporting parents, while the federal government lies to the public, quotes the Bible and relentlessly ignores a judicial order to reunite. Morbid symptoms indeed. Family members are "processed," mothers are told "let me take your child for a bath," and then they are flung across the nation—a diaspora of bodies branded by state violence—stuffed into the bowels of distinct bureaucratic departments, viciously and ironically named Homeland Security or Health and Human Services. Of course, the data bases are not aligned. Children's bodies and records evaporate; mothers' ankles swell from too tight ankle bracelets; wealthy "soccer moms" join with domestic workers in New York City to send bail money for a Guatemalan mother locked up in Arizona so she might travel—in a synchronized caravan of drivers—across the country to reunite, perhaps, with her babies. As of this morning, many parents are being deported and/or told they may be reunited with their children *only if they go home*.

Far less attention has been paid to the "developer" friends of Trump, the CEOs of private prisons and corporations who are being handsomely and outrageously subsidized with substantial tax dollars, to house, feed and transport these families. The Guardian (2017) reports:

> Individual investors in US private prisons are poised to collect their most lucrative earnings ever thanks to changes in the tax code signed by Donald Trump, continuing what has been a banner year for the industry since the 2016 election (Lartly, 2017, www.theguardian.com).

Witnessing this upward redistribution of tax dollars, the racialized containment and splitting of families, and the public displays of sadism, we are all ensnared: as witnesses, perpetrators and victims, of course with differential keloids rising on our souls and our nightmares. Cheaper, "evidence based" alternatives to mass incarceration have been dismantled, replaced by costly, inhumane strategies of public state sponsored abuse:

> The Family Case Management Program, allowed families to be released and receive intensive case supervision; help with child care and education; connections to legal counsel;…. The program was 99 percent effective at having families show up for check-ins … cost of only $36 per day per family, it was far cheaper than family detention, which costs over $300 per person per day. The Trump administration shut down the program last June… Wolgin, 2018, www.americanprogress.org/issues/immigration/news/2018/06/20/452571/incarcerating-entire-families-cannot-solution-separation-children/.

This contemporary enactment of State Violence and Public Sadism flows from a long shameful line of state sponsored child snatching from families of color. While it is always tricky to map parallels and continuities across time and place, it is equally dangerous to assume this form of structural violence is new or exceptional (Bhatia, 2017; Foucault, 1972; Narayan, 1998). The brutal tearing of brown babies from their mothers/fathers/siblings was constitutive to the brutality and profit of slavery and Black child selling; at the heart of "kill the Native, save the child" rhetoric that justified the Indian boarding schools; echoes the coercive internment of Japanese American children and families during World War II; resonates with the contemporary systematic removal of Black and Brown children and disparate placement into foster homes, and aligns—in purpose, design and consequence—with the over-incarceration of Black and Brown adolescents, and their ruthless murder by police. While state violence against communities and families of color has been a national—if heterogeneously enacted—tradition in the U.S. the public performance of State Sponsored Sadism—torturing children wrapped in mylar, living in cages, sobbing for their mothers in native tongues—seems somehow new, perhaps "modern."

And in the midst, some of us ask about the damage to the children and families and also who benefits, what (de)forms of national values/ethics/identities are circulating and being "naturalized", what are the consequences for those most affected and also for those who bear witness in a passive complicity of paralyzed helplessness. Activist scholars considering how to craft research from rage and in solidarity, that is response-able to, and with, communities under siege while also revealing the entangled links between state violence, historic echoes, neurological damage and the toxicity of enforced compliance in the bodies of children separated and parents, and also in border guards, precariously "documented" people living in the U.S., teachers, social workers, health care providers, youth workers and those of us whose families immigrated here, sailing past the Statue of Liberty at the moment when our country needed/wanted/welcomed white skinned immigrants with little education, who didn't speak English, who were demonized and then naturalized and assimilated.

This chapter offers a minor contribution—a non-linear, selective and strategic history of critical social psychological projects, hatched across the twentieth century, by U.S. scholar-activists who refused to fetishize "symptoms" of oppression, and insisted instead on working in the borderlands between research and social movements for justice. My gift to readers: Many of these texts sketched in brief below have been extinguished from mainstream psychology. These activist scholars and texts have suffered "death by canon," whited-out from the imperial, positivist, and universal-law seeking/power flattening history of psychology, through acts of epistemicide (de Sousa Santos, 2014). With a sense of urgency, we excavate epistemological, ethical and relational commitments to community struggle as we unearth critical theoretical frames, a wide range of methods and wild connections forged by scholars with(in) prisons/unions/activists and public intellectuals through capillaries of popular dissemination: arts/music/theatre/pageantry/satire/journalism.

My self-critique: the archive I serve up is both delicious and a bit random; this is a rhizomic gathering of activist scholars and fugitive sites. I make no claims to comprehensiveness but want to introduce you to those crazy aunts and uncles exiled from the family disciplinary bible (see: Adams, Dobles, Gomez, Kurtis, & Molina, 2015; Bhatia, 2017; de Sousa Santos, 2014; Montero & Sonn, 2009; WITS Fallists,

2017). Suturing these ghostly ancestors and texts back into our disciplinary history—hopefully without sounding either romantic or "Whiggish"—in this chapter we will confront the joy, messiness and contradictions of critical science making, wander into the creative margins of disciplinary history (re)making, taste the desire of activist-scholars to "be of use" in revolution making and we will end reading a short piece, *Crying Over Spilt Milk*, drafted in the parking lot of an ICE detention camp in 2018 so we may collectively embody the affective weight of *living through* as others' bodies fall.

Borderlands: Critical PAR Spaces for Democratizing Knowledge

The poet, lesbian feminist, Chicana writer Gloria Anzaldua lends us the language of borderlands, to conceptualize spaces where blood, oppression and violence spill over, even as new identities, knowledges and radical possibilities metabolize:

> The U.S-Mexican border es una herida abierta where the Third World grates against the first and bleeds. And before a scab forms it hemorrhages again, the lifeblood of two worlds merging to form a third country—a border culture.
>
> Borders are set up to define the places that are safe and unsafe, to distinguish us from them. A border is a dividing line, a narrow strip along a steep edge. A borderland is a vague and undetermined place created by the emotional residue of an unnatural boundary. It is in a constant state of transition. The prohibited and forbidden are its inhabitants. (Anzaldúa, 1987, p. 32).

Over time and space, a series of research collectives have been hatched in border spaces throughout the U.S.—settlement houses and prisons, in underground meeting spaces, community centers and universities—producing white papers, class action lawsuits, pageants, popular education, music, art, satire, radio schools and social theory. In these spaces, knowledge production has been democratized (unevenly and with contradiction); pitched toward collective action, and the footprints of oppression have been traced right alongside the unjust accumulation of power; designed to contest lynchings and denial of human rights, corporate greed and environmental racism, state violence and aggressive policing, white supremacy and educational injustice, and the vast archipelago of racialized incarceration and disenfranchisement—all horrific and well established traditions in the United States.

The projects reviewed in this essay are organized through a simple question that lies at the heart of the decolonizing movement—to whom are we accountable? (Patel, 2016; Segalo, 2011). These scholars, to date, have been inadequately recognized in the legacy of psychology. I name them not to individualize or reify them as "protagonists," but to recognize scholars of color, white women and leftists excised from the canon. They sit alongside each other as a curated montage; the shoulders upon which those of us engaged in decolonizing work have been strengthened by the writings of contemporary critical scholars—some first cousins to community psychology: Linda Tuhiwai Smith (2005), Nelson Maldonado-Torres (2016), Leigh

Patel (2016), Stevens, Duncan, and Hook (2013), Anzaldúa (1987), Audre Lorde (1984) and James Baldwin (2017). These social scientists/activists/writers have massaged/stretched/punctured the membrane of the academy and community, positioning themselves within/alongside movements for justice, believing themselves accountable to "the people" whose stories, knowledge, critique, aches, struggles, joys and freedom dreams had been smothered (see Table 1). Igniting staccato moments of epistemic rupture, not necessarily aligned over time or space, these writers have been willing to call out power, contest privilege and profit, and widen the gaze to include but not fetishize the "damage" endured by communities under siege.

Refusing to Fetishize Symptoms/Insisting on Analyzing Power, Structure and History

In the late 1890s, in Philadelphia, W. E. B. Du Bois was asked by Philadelphia's Susan Wharton to investigate "the Negro problem." Understanding that the invitation was framed in a way that focused problematically on the Black community as the *source* of social ills, Du Bois was determined to analyze the Negro problem as a "symptom, not a cause," of the troubling economic and racial order. He undertook the systematic analysis of high morbidity rates among Negros, comparing them to Whites in North, South, Europe and Negroes in Europe 1884–1890 by interrogating history, policies, opportunities, housing, food, sanitation, access to medical care. Du Bois entered the politically vexed research project with full knowledge that evidence of historic oppression was being (mis)cast as biological weakness and/or cultural depravity. Aware that he had been hired to document the pathology of the Negro, Du Bois (1899) calibrated the economic, historic, educational and political groundings of these problems reversing the gaze of causality that landed squarely and viciously on the ideology of Negro inferiority.

While *The Philadelphia Negro*, and *The Souls of Black Folks*, are recognized by many as classic sociological texts, the academy was never a welcoming or comfortable space for Du Bois and for decades his work was neglected within academic sociology. At Penn, he was not allowed to teach White students; he was only rarely—if ever—cited by his white mentors/colleagues and his legacy has been severely minimized within sociology (see Rebaka, 2010). He ultimately left university life concerned that the casual timing and the limited reach of academic scholarship were insufficient to contest the accelerated evidence of lynchings. While much of his early work was ignored by white academics and roundly critiqued by scholars of color, as elitist (e.g. notion of the Talented Tenth), later in life the tenor of his work radicalized, influenced greatly by the political/intellectual possibilities opened by Marxism.

In the same historic era, sociologist Jane Addams and her partner Ellen Starr established Hull House, in 1889 in Chicago, as the first settlement house in the

Table 1 To whom are we accountable

Accountable to whom/for what?	The object of contestation	Researchers	Modes of knowledge production and performance
Accountable to historically marginalized communities of "Negroes" and immigrants Late 1800s/early 1900s	Systematic documentation of the downstream consequences of racist/classist policies on poor Negroes and immigrant women	WEB Du Bois Jane Addams	Journalism, social theory, political activism in and beyond U.S., pageantry and theatre Political/Civic engagement, advocacy, national organizing, drama
Accountable to contesting White supremacy 1930s–1950s	Through systematic gathering of statistical data, stories and parody, these scholars challenged limits on voting rights, civic participation and the prevailing use of IQ tests to sustain White supremacy	Guy Benton Johnson Horace Mann Bond	Technical reports, participation in The American Dilemma and newspaper/popular articles to the white community (1930s) challenging the benefits of supremacy Publishing largely in Crisis magazine, and collaborating with lawyers, Bond posed a serious critique of testing as a tool of racist oppression
Accountable to documenting the abuse of state violence, imperialism and racialized criminalization 1980s	Contesting the "dominant lies" and chronicling the systematic oppression of workers, farmers, indigenous people and people who have been incarcerated	Jean Paul Sartre and Franz Fanon Ignacio Martin Baro Brinton Lykes GreenHaven Think Tank	In the short-lived preface, Sartre challenges the French to recognize and account for the consequences of colonialism Deliberately in Spanish only The People's Census Center for New Leadership—"penealogical glossary and analysis"
Accountable to local movements and public science for the collective good 2000–forward	Mobilizing everyday people, activists and academics to create policy counter narratives from the ground on racial, economic, educational and criminal justice oppression	The Public Science Project	

United States, for recently arrived immigrants living collectively with Addams's and Starr's elite friends. In the Hull House, social inquiry, policy research and political debate were facilitated across class lives, as residents of Hull House and the professional researchers working with them collaboratively investigated truancy, typhoid, midwifery, housing, garbage collection, and cocaine use, publishing again scholarly articles, popular education materials, writing essays for newspapers, and like Du Bois, they launched a drama club and theater group for radically re-imagining the praxis of "dissemination."

Hull House staffers not only studied Chicago's social issues, but they immersed themselves in the Chicago and Illinois politics. Addams became the garbage commissioner of the city's 19th ward and was embroiled in one of the great labor disputes of her time, the Pullman Strike of 1894. In writing of the strike, Addams made clear Hull House's unequivocal support of trade unionism. Always speaking back to power and with communities under siege, she was explicit that capitalism conflicted with her group's "ethical demands" not only for "social righteousness but for social order." In 1910, Addams delivered a renowned address, Charity and Social Justice, to the major philanthropists in Chicago, accusing them of *creating* the very problems and inequalities they were presumably committed to *resolving*. She adamantly distinguished "benevolence" and "justice." As time passed, Addams became evermore involved with pacifist causes as well as social and feminist issues. In 1909 she became a charter member of the NAACP. Dedicated to weaving politics, social work and research, she was named the first Vice President of the National American Women's Suffrage Association 2 years later, became Chair of the Women's Peace Party, and later the first president of the Women's International League for Peace and Freedom (WILPF) until 1929 (2002).

Contesting white supremacy. The next decade, in the U.S. segregated South, in the midst of the rise of the KKK, Guy Benton Johnson, was a white leftist sociologist committed to documenting racial disparities in attitudes and opportunities and writing *to whites* about the pathology of white supremacy. In January of 1934, on the faculty of the University of North Carolina, Johnson drafted a letter to the Commission on Interracial Cooperation, seeking $2500 for a Study of the Participation of Negroes in the Life of the South in varied orbits of social life: education, economics, health, crime, family life, moral life and everyday lifestyles. In a dusty archive at the UNC library, one finds a folder marked Study of Participation, and a set of hand-written notes to self: "white supremacy of the prisons" and "see about getting Du Bois on the committee" and "get rates of Negro access to loans, crops in certain states" (archive reviewed 2017).

Johnson (2018) was explicitly interested in confronting white southerners with a simple question: is white supremacy even beneficial to Whites? In his essay "Does the South owe the Negro a New Deal? Johnson characterized the South as a "deficiency area"; he opines, "One might raise the question here as to whether the South's preoccupation with "white supremacy" has not served to retard the progress not only of the Negro but of the white man and of the whole region as well." He elaborates on the "stagnation of political thought and leadership; retardation of economic progress; deadening of community consciousness and moral sentiments as a consequence of rationalizing and defending the policy of Negro non-participation." Not at all naïve to the white resistance fomenting in response to emancipation and talk of "Negro" voting rights, the right to serve on a jury, be hired as a police officer.... in his outline for the report he authors a surprisingly empathic (to whites resisting African American voting rights) section called "The Difficulties":

> "The difficulties are so numerous as to frighten all but the excessively courageous—and perhaps the fools who rush in where angels fear to tread. There are fears, obsessions, phobias; there is the inevitable avalanche of questions concerning the social sphere—Do you want to be arrested by a Negro? Do you want to eat with a Negro? Do you want your daughter to marry a Negro?—and there is the whole complex of vested political and economic interests". And then he asks his presumably white audience if they seek to further alienate and suggests the possibility of radicalizing the Negro: "can the South afford to continue the old policy toward the Negro? Does it wish to alienate the Negro until, embittered and resentful, he turns to radicalism as his helpmate in his struggle for rights?"
>
> While Johnson was speaking largely to the white community about white supremacy, Horace Mann Bond, a Black psychologist and educator writing through the 20s through the 1960s in Atlanta Georgia, was publishing satirical parodies on white supremacy in the IQ/ testing movement that was gaining momentum; he wrote, primarily, for and with the Black community, with his first publications in the NAACP magazine The Crisis in 1924. Adept at critical counter-analyses of statistical matter, Bond critiqued the logic behind Brigham's then quite popular conclusions that the lower African-American test scores on the Army intelligence tests indicated an inherent intellectual inferiority of the Negro race. Bond concluded that "the medial score of White soldiers from the states of Mississippi, Kentucky, Arkansas, and Georgia, averaged. …the mental age of a twelve and a half year old child" And he asked "Are the exponents of intelligence tests as discriminators of racial differences prepared to assert that the white population of Arkansas is inherently and racially inferior to the whites of another section of the country?" (p. 200).

In the mid-1950s when a group of White Southern Senators signed the Southern Manifesto in opposition to racial integration and the Brown vs. Board of Education decision, arguing that African Americans were not sufficiently intelligent to participate in the same schools as Whites, Bond published a parody of their arguments in an essay entitled "Intelligence of Congressmen Who Signed the 'Southern Manifesto' as Measured by IQ Tests," calculating the average intelligence of signing senators was in the lowest 20% of American Whites, who attended colleges of the lowest 10%of median National scores (Jackson, 2004).

By 1959, with sharpened empirical evidence and wit, Bond suggested to Supreme Court Justice Thurgood Marshall that the government could save money on IQ testing of all children and just count the number of toilets in a household to determine so called "talent"—more toilets, more talent; fewer toilets, less talent. He recommended that instead of administering IQ tests to youngsters, the government should 'subsidize the building of two-holers, three-holers or even eight-holers.' (Bond, 1959b, p. 9 in Jackson) That's when Marshall started calling him "Mr. Toilets." Drawing on empirical creativity and political satire, Johnson and Bond were bold and clever, nimble with evidence and persuasive rhetoric. Writing for distinct audiences, their motives were linked revealing the pathology, hypocrisy and dangers of white supremacy. Both worked out of and through the academy to advance arguments most unpopular in their times, and in the South.

Contesting the dominant lies of colonialism and state violence. Our journey moves into the 1960s, sampling "tapas" of critical scholarship penned to expose of the violent appetites of colonialism abroad, and oppression at home. In the opening pages of his short-lived preface to the original volume of Frantz Fanon's *Wretched of the Earth*, Jean Paul Sartre (1961) writes to the French: *"It is you who are zombies"*.[1] Confronting White Europeans with complicity in birthing scars, chains and trauma throughout the African continent, Sartre challenges white silence, denial and the systematic turning away from the bloody footprints of colonialism.

The preface, of course, precedes Fanon's brilliant and highly circulated text and was removed soon after original publication. But while the twinned texts existed, linking arms, emboldened, between the same covers, they were directed to two distinct audiences. Sartre opens his preface with an attack on the colonial project in Algeria:

> The European elite undertook to manufacture a native elite. They picked out promising adolescents; they branded them, as with a red hot iron, with the principles of western culture, they stuffed their mouths full with high sounding phrases, grand glutinous words that stuck to the teeth. These walking lies had nothing left to say to their brothers; they only echoed... it was a golden age.

He then implores his fellow Europeans:

> Europeans, you must open this book and enter into it. After a few steps in the darkness you will see strangers gathered around a fire; come close and listen for they are talking of the destiny they will mete out to your trading centres and to the hired soldiers who defend them. They will see you perhaps but they will go on talking among themselves, without even lowering their voices. This indifference strikes home: their fathers, shadowy creatures, your creatures were but dead souls; you it was who allowed them glimpses of light, to you only did they dare speak and you did not bother to reply to such zombies. Their sons ignore you; a fire warms them and sheds light around them and you have not lit it. Now at a respectful distance, it is you who will feel furtive ... and perished with cold. Turn and Turn about; in these shows from whence a new dawn will break, it is you who are zombies....

As if an echo of Du Bois, Sartre reverses the gaze, as he contests the psychiatric/psychological misrepresentation of French colonialism as if it were inherent to Algerian pathology.

Sartre accuses his French readers of complicity:

> It is not right, my fellow countrymen, you who know very well all the crimes committed in our name. It's not at all right that you do not breathe a word about them to anyone not even to your own soul, for fear of having to stand in judgment on yourself. I am willing to believe that at the beginning you did not realize what was happening; later you doubted whether such things could be true; but now you know, and still hold your tongues. Eight years of silence; what degradation! And your silence is all of no avail; today the years of silence; what degradation!

Fanon wrote for/to Algerians; Sartre to the French. For a short few years, they performed a tragic duet, as if both on stage, speaking to different audiences, hitting

[1] Fanon's widow removed Sartre's preface in 1968 given his endorsement of the Zionist project in Israel.

a few shared notes, a jazz of descriptive narrative, reversals, rage, comedy, dystopic imaginaries and calls for action.

More than a decade later, in 1982, social psychologist and Jesuit priest Ignacio Martín-Baró returned to El Salvador, from graduate School at the University of Chicago to initiate a series of "the people's" research projects at Universidad Centroamericana. Motivated by the vibrant intersections of liberation theology and liberation psychology and working as the Director of the University Institute for Public Opinion, "Nacho" argued for participatory research, of and by "the people," as essential to "challenge the official lies" of the dictatorship. In his short career, Martín-Baró seeded and launched a line of liberation psychology, by and for the poorest peasants, farmers, fisherman and indigenous peoples of El Salvador. In November 1989, he and five colleagues were murdered by a counterinsurgency unit of the Salvadoran government elite.

While Sartre and Johnson chose to speak back to white Americans and Europeans and Adams to white elite philanthropists, Du Bois, Bond and Martin-Baró were committed to educating their own communities, piercing the ideological anesthesia and hoping to awaken a sense of injustice. For Martin-Baró this meant publishing only in Spanish for his Latin American communities, speaking with oppressed communities to "de-ideologize" experience and mobilize action. In the 1990s, a set of Boston-based psychologists recognized the need for North American psychologists to be educated by Martin-Baró's writings, and organized to translate his brilliant and inspiring essays, available in *Writings for a Liberation Psychology: Ignacio Martín-Baró* (1994; see the ground breaking work of Brinton Lykes (1997) with Mayan activists in Guatemala for instances of liberatory research with Central American communities.

During this same era, in the late 1980s, the Center for NuLeadership—an activist research collective of incarcerated men—coalesced within the Greenhaven Think Tank, at Greenhaven prison in New York State. Led by former Black Panther Eddie Ellis, the Center was dedicated to re-writing the story of racialized mass incarceration from the perspective of those who knew best—incarcerated men of color. Think Tank sought to shift language (rejecting "inmate," "corrections" and "superintendents"), to investigate and ultimately prevent the rising numbers of Black and Latino men consigned to the New York State prison system. Under the direction of Ellis, who was imprisoned for 23 years, and with the help of psychologist Kenneth Clark, the men of the Center—"street penologists"—designed a study and determined that 85% of the New York state prisoners were Black and Latino, and that 75% of them originated in seven neighborhoods downstate. From within prison, members of the Blacks' Resurrection Study group and the Latino-based group Conciencia integrated the Seven Neighborhood results into a Greenhaven Think Tank (1997) policy document. They urged a nontraditional policy analysis of the seven "symbiotic neighborhoods"—Lower East Side, South Bronx, Harlem, Brownsville, Bed/Sty, East New York, and South Jamaica—to the then 62 state prisons, and recommended a nontraditional vision: that these men, while in prison, should be trained in community development, mentorship, and adult education and once paroled back to those communities, should be funded to help rebuild the community through internships, mentoring, and community programs. The men even detailed their vision for a prisoner-run model prison. While Ellis passed away recently; the center survives in Brooklyn, and the soul of the center, like much of the leadership in the seven neighborhoods, can be found in cohorts of men and women formerly involved in what they call the "criminal punishment" system, conducting research, training police, facilitating healing justice work with local youth and advocating for policy change.

The Public Science Project: A participatory policy collective. For the past 25 years, at the Public Science Project (PSP), participatory policy projects have been designed by collectives of researchers, activists and everyday affected people on key issues of educational, racial and criminal (in)justice, in prisons, schools and under-invested/over-criminalized communities, tracing neoliberalism and the privatization of education, racial capitalism and aggressive policing, gentrification, gender/sexuality violence from the State and from loved ones. Theoretically and historically grounded, the projects blend quantitative and qualitative methods, often woven with popular education, performance, litigation and the arts.

We are institutionally housed within Critical Psychology at the Graduate Center, CUNY. The projects are rooted in a design commitment to what Jessica Ruglis and I have called *circuits of dispossession, privilege and resistance* (Fine & Ruglis, 2009). And we routinely confront, within these quite diverse contact zones, what Gloria Anzaldua would call "choques"—difficult eruptions at the membranes of power, privilege, race, experience and trauma. While never fun, these choques erupt usually in ways that provoke a radical redesign of the work, unleashing new knowledges and often deeper trust, at least more insight/incite, within our research collectives.

Our projects are organized loosely around a set of shared commitments:

- We understand critical PAR to be an epistemology, not a methodology.
- We contest traditional definitions of "expertise" and "objectivity" and gather intentionally very different kinds of knowledges/experts/experiences
- Our work is led theoretically, empirically and analytically by those who have most intimately experienced the wounds of injustice, working in what we call "contact zones" and dedicated to exploring rough spots of disagreement, dissent or what Anzaldua would call "choques."
- We aim toward Sandra Harding (1994) would call "strong objectivity" and contest what Ignacio Martín-Baró (1994) called the dominant lies. Our designs document downstream "symptoms," but also trace back those outcomes through circuits of dispossession, privilege and resistance.
- And our first "products" are collectively designed to widen the radical imagination of social possibilities, and to "be of use" to the community most adversely affected by injustice (e.g., Payne, 2013; Segalo, 2011).

The Morris Justice Project on aggressive policing. Jackie Robinson, Fawn Bracy and Nadine Sheppard knew well the tactics of aggressive policing in the 44th precinct. Since their boys entered early adolescence, they had been trading photos of police brutality from the varied floors in their apartment buildings; archiving and analyzing the patterns; following each other/their friends to the police station to "prove" their sons' innocence. After years of community-based documentation, and after a few months of working with a lawyer, they approached PSP seeking assistance to develop a more systematic community-wide survey. Brett Stoudt and Maria Elena Torre joined the research team (see Stoudt et al., 2015) with an expectation of working together for 6 months.... It's been 5 years and counting.

The Morris Justice Project, a deeply rooted community survey of aggressive policing in a 40-square-block area of the Bronx, gathered empirical evidence from secondary of NYPD data and then community generated and analyzed surveys of more than 1000 residents on over-policing and the consequences for children, community safety, and democracy. The findings were introduced as legal evidence into the stop-and-frisk *Floyd v. City of New York* case, have been shared with the Coalition for Fair Policing, and have been replicated by young people in the community group Make the Road in Brooklyn. The research team has presented their findings at the White House conference on Citizen Science, to police departments in Toronto and Paris, and at a variety of academic/organizing/political meetings. They shared their work throughout the Bronx and Brooklyn, with the mayor's office, and in advocacy campaigns. As a team they facilitate Stats 'N Action workshops at the Yankee Tavern for community people to get familiar with local policing statistics, hold Know Your Rights sessions, and distribute cleverly designed fliers of their findings at Yankee Stadium. The T-shirts they distribute are tagged with open-ended survey responses that read, "It's not a crime to be who I am" or "Why do I always fit the description?" The shirt is most sobering in child's size Small. As the Morris Justice Project (MJP) website reveals (see: www.morrisjusticeproject.com) the project generates provocative legal texts, videos, local art work and pamphlets (see Fig. 1 below) on racialized policing, and is now involved in a deep community study of data to generate radical alternatives for community safety.

While the MJP has been a powerful enactment of activist public scholarship between university and community, effective legal testimony and radically thrilling inclusion of the arts, the long and painful history of aggressive policing in New York City reflects a stubborn national tradition, always contested and yet relentless. And now, with a strong sense of "NO Research or Policy On Us Without Us," these practices are under siege by a robust collective forged between academy and neighborhood, across Black Lives Matter, neighborhood organizing, Muslim American struggles, LGBTQ organizations, and movements for immigrant justice including the rights of undocumented New Yorkers. This is not a progress narrative—where critical evidence slays the evils of racialized policing. Critical PAR is one more tool in movement work, policy advocacy and leadership building, but the abolition of racialized state violence is not an empirical problem.

Across the projects just described, we trace a deep strain and long staccato history of critical social scientists working with everyday people and movements in epistemic borderlands, generating new forms of knowledge about oppression and privilege and producing a range of provocative, empirically grounded materials for the courts, theatre, popular education, social theory, and for community organizing.

A Tale of Two Neighborhoods

Using 2011 NYPD Stop, Question, and Frisk data and 2010 US CENSUS DATA

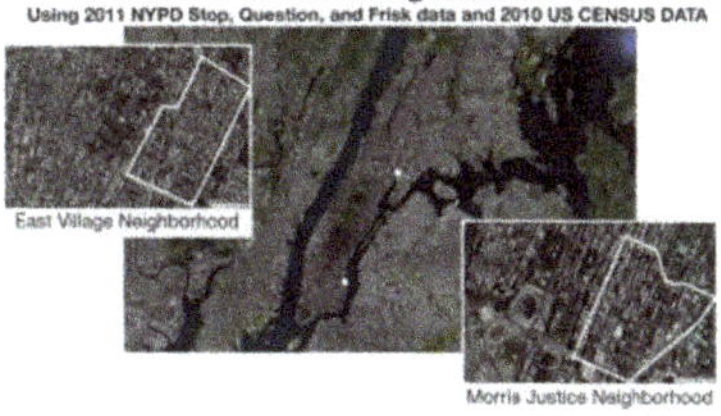

Residents in the Morris Avenue neighborhood are policed differently than residents in a section the same square miles and similar population size in the East Village. In fact, *The New York Times* reported that our precinct (the 44th) and the neighboring 46th "use physical force far more often than the police do anywhere else in the city" (August 15th 2012). **A comparison of how our 40 blocks just east of Yankee stadium are policed versus a similar area in the East Village just east of NYU shows how unequal things are depending on where you live and who you are.**

East Village Neighborhood (.243 sq mi. Pop. 22,889) Part of 9th precinct 68% White		**Morris Justice Project Neighborhood** (.239 sq mi. Pop. 19,173) Part of 44th precinct 96% Black/Latino(a)
2135	Total police stops	4882
45.5% (971)	Stops involving frisks	80.1% (3909)
12.7% (271)	Stops involving searches	14.6% (711)
81.9% (1748)	Stops NOT leading to arrest or summons (innocent)	92.5% (4517)
.23% (7)	Stops leading to gun recovery	.12% (8)
1.8% (38)	Stops leading to knives, cutting instruments, and other weapons recovered	.82% (40)
6.1% (130)	Stops leading to contraband recovery	1.5% (75)
24.3% (519)	Stops involving physical force	58.6% (2859)
68.8% (357)	Percent of physical force stops not leading to arrest or summons (innocent)	90.8% (2597)
15.8% (338)	Stops leading to arrests	5.4% (266)
2.3% (50)	Stops leading to summons	2.4% (118)
16.5% (352)	Stops on people 20 years old or younger	42.2% (2066)

Fig. 1 The Morris Justice Project. http://morrisjustice.org/report[morrisjustice.org]

In each setting, crucial questions of power, respect, vulnerability and disposability emerge. In many of these cases, the stakes were quite high for people in community, who continue to be harassed, threatened and taken off the street, even as they would watch their children endure the same. And some of the academics paid a severe price of prison, exile, loss of job and, in the case of Martin-Baró, loss of life. And yet today in these "revolting times," it seems that the commitment to solidarity policy research, even among psychologists, may be rising—if the numbers of applications to our Summer Institutes on Critical Participatory Action Research are any measure.

Today, again, we all—across national borders—find ourselves saturated in morbid symptoms, trying to figure out how to be of use in the *throbbing present*. To close this essay I want to escort readers into one more borderland—where varied bodies and tongues come together, from around the world, held in sadistic liminality, between nations, dangling at the membrane of life and death, well financed by the state and profitable for birthing a new industry of private detention center management corporations. This is a dystopic borderland where bodies, histories, subjectivities and intimate knowledge of state violence mingle, where affects of rage, despair and solidarity materialize. In this moment it is hard to know how, in the end, the snatched babies and mommas, and the rest of us will fare. And yet, at this moment, before a narrative of clarity sets in, it seems crucial to bear witness to a chaotic grotesque borderspace, where new knowledges of evil, (in)humanity and survivance, terror and terrorism from above, are being borne.

I entered simply as a psychologist scribe, a volunteer with Human Rights First, willing to interview "detainees" about health conditions and human rights, at the Hudson County Correctional Center, February 2018. In this space, all our bodies were penetrated by dynamics of state initiated terror—but some of us were free to go home.

Crying Over Spilt Milk

Michelle: "This might be an awkward question, but how do you handle sexual violence within the facility?"

As he turned to me, Sargeant McCoy seemed startled; he had been accompanying the Human Rights First monitors for 4 h, attentive to our questions, hugely responsive.

"You mean, like PREA (Prison Rape Elimination Act)? An officer and a detainee?"

"well, yeah."

"They—the detainees—go right to the KIOSK and submit a grievance. Shouldn't come thru a guy in uniform, goes right to the county. If someone comes to me, I say put it in the KIOSK. That's consistent with 2008 Standards."

This was the third time I heard about the KIOSK, and so I asked to see one. "When we are in the pod, I'll show you."

We traveled up to the sixth floor, and then down to the third; to examine the pods of men and women "detainees." When asked by a lawyer, also on the Human Rights First monitoring tour, about visits and their aftermath, McCoy explained: "Well, yeah, they have to be strip searched after a visit. For everyone's protection."

We visited the "medical suite"—and thru the plexiglass we saw the "observation" housing cells, where "we place people on suicide watch for their own protection." In the old days we called them Solitary Confinement.

The medical doctor who was with us asked, "How do you know when they are ready to come out?" "They're in 23 hours, and then out for an hour. Someone from medical evaluates them. Could be 24, 48, 72, depending on what they need."

We all knew that the facility has suffered headlines due to a number of suicides and recent deaths. That's how we got in.

Our task was to interview as many "detainees" as we could to gather information about health care, human rights and the needs of the detainees while in "state protection." We divided into those of us who could speak Spanish well enough to facilitate an interview; French, Creole and English.

I didn't trust my Spanish; I can fake it, but felt obligated to carry these stories out with dignity. So, I met with Tina, and then Rodrigo.

On the third floor I met Tina, code switching—all in English—from a discourse of yearning mother to outraged captive, originally from an island in the Caribbean. "I'm 40 and came here to the U.S. when I was 28." "Why? Nothing dramatic. A better life" A beautiful black woman, small, stunned, eloquent, narrating in a split screen. The soft maternal alternated with the traumatized voice and then they were one: "last Thursday I was breast feeding my baby, in my pajamas. He's two next Tuesday. He has asthma and I was feeding him before school. My older boy was getting ready for school—he goes to an all boys academy in Brooklyn. Waiting to hear about college. There was a knock on the door. There were three ICE officers, but the one who spoke, who lied, a woman. I opened the door. "The judge wants to see you, just for a few minutes. There's a harassment charge. You will be back by noon. Bring some money, Metro Card." I explained, "You can see I am feeding the baby, I am not dressed. I can come down to the court later this morning." "Let your husband take care of the kids and you'll be back soon. Just a few minutes. You'll be back by noon."

"I followed them, I didn't want trouble, not in front of the children."

"They took me to Brooklyn court, then Van Wyck, and then—handcuffed, shackled with a big metal belt around my waist, they put me in a bus with 12 men. I was so frightened, so frightened.' "When they picked me up the woman ICE officer kept saying why is she so angry?" "And now I am here. I don't know why, I don't know for how long, this is hell, it's fuckin' hell. Get me out of this fuckin' hell. There is drama all night; 60 women; one micro-wave. I don't mean to be racist but the Hispanics, they love the chicken, look [she points to the enclosed triangle of 'outdoor' space, that is fenced in, off the third floor]. They leave their chicken there overnight. And they sing and dance all night. Just too much drama. And during the day—They just shut off the phones, they lock them down from 1 to 3. I was on the phone with my husband, he's trying to find a lawyer but he's out in Brooklyn and I am here. I can't eat this food. I don't know why I am here, how long, my husband doesn't know where Kearney is, can't get a lawyer. I am ready for voluntary deportation, send me back. This is hell, or I will kill myself."

I reached for her arm, touched her gently and whispered, "if you really feel suicidal, let's talk. if you don't, please don't say it. They will put you in solitary. I just saw the cells, you don't want to be put in there."

The image of this beautiful strong mother, talking to me about her son "He wants to go to SUNY Buffalo, but maybe if he goes to CUNY you will teach him…" and her eyes drift away. She smiles for the first time. I ask the names of both boys, Dwayne and Duquan, which provokes another guilty memory. "I hope my husband found the nebulizer." I envision her in the "observation cells downstairs," cracking, folding, broken in solitary. Sargeant McCoy interrupted, "So you were asking about the KIOSK. That's it. Against the wall."

I turned to see a screen, no chair, accessible to all—and therefore visible to all. I didn't approach until we got to the men's floor.

Tina continued, "That kiosk is a money machine. It sucks your money. They give you one towel, so I bought another. I need sweat pants, $13.95; asked for a toothbrush, they said didn't Friends and Family give to you? I said no or I wouldn't ask. Buy it. So I had to buy a toothbrush, mouthwash, ramen noodles, cant' eat the frozen waffles or white bread and bologna. The Spanish women they love the chicken, they keep it out there—pointing to the small space they call the "rec area" a triangular space surrounded on one side by plexiglass, another by a kind of barbed fence; probably 30 × 20, where you can "exercise." "The Spanish women put the extra chicken there over night. I will lose 20 pounds here, I am not eating that mystery meat." "No one gives a shit about us."

"There's one nice guy, from ICE, that's him; one of the staff of color." "County" seems to be white, and ICE people of color, "he told me that immigration court is backed up—then why pick me up, they say 3–4 weeks, but no one knows. I just want to see a judge and get a voluntary deportation." Then she switches to momma mode, "taking my children, not leavin' them here. but my son wants to go to college. Maybe you can teach him if he's at CUNY. He just decided to drop physics and advanced calc = I figured e's a senior, that's ok. He's already applied to schools." And then body and voice returned to form of retreat, the voice of despair, a soft, bent resignation blanketed over a fierce momma body, ready to go home to "my mom and dad, they still live in the Caribbean." Her husband set up a three way call from Brooklyn and Kearney to her parents, "my mother was heart broken to hear me in here."

Tina asked me for nothing. She was most incensed by the brutal banal details of this body snatching, family disruption, imperial grab: that a Black woman officer came to the door and demanded she leave; that "they lied"; that "ICE is the major human trafficking organization in the world"; that she has to pay taxes into the KIOSK; that she is dangling between countries, states and communities, swinging between freedom and State containment, without her children or a lawyer. And I sealed her lips. She couldn't even say suicide without more restrictions imposed on her willful: breaking soul.

The women's pod—I learned as I traded notes with the other Human Rights watch monitors—was somnolescent; sleepy. Maybe 30 of the women were in their beds. One of my colleagues—a lawyer—suspected they were given "meds" to sleep. Peace in the facility requires that the women were drugged. And when they go home—wherever home is—the yearning for meds would intensify. "I pump my breasts in the shower, I am still producing milk but worried about the chemicals in the air." Tina whispered The floor smelled like ammonia and old chicken, and aching momma hearts.

Tina interrupted our interview to go to the bathroom, to squeeze the drips of remaining milk from her breasts.

When she returned, our eyes met, and knowing we were leaving, that her story would be archived in the long arc toward justice, I drifted off, filled with shame at our country and that I didn't know; that I had no idea.

During our conversation, in the background, a woman in orange, who had recently arrived from China sobbed loudly and uncontrollably. Sargeant Coyne and another white officer were trying to calm her; she was surrounded by women from Korea, Turkey, Uzbekistan, Trinidad, rubbing her arms, trying to calm her, across linguistic walls. The two officers were chatting, loudly, "think she's suicidal?" "should we bring her down?" Coyne, in another rupture of what seemed like empathy in hell, turned to the monitored team, "I just broke every rule in the book, I downloaded a translation app, but still can't really understand." More women circled around, as if in a medieval ritual of women blanketing each other, conversing with eyes and hands, knowing only Spanish, Korean, an African American woman, huddling around her. And she cried. They decided not to send her down. Coyne admitted to me, "I think she was upset she would miss Chinese New Year in Long

Island." Relieved this woman wouldn't be placed in solidary, I assured him he was probably right.

When I got home I read the NPR special on human rights violations in NJ detention camps:

> "NPR, 12.15.17 immigrants detained at four large centers used by Immigration and Customs Enforcement are subject to inhumane treatment, given insufficient hygiene supplies and medical care and provided potentially unsafe food, according to a federal report [Inspector General's Office of the Department of Homeland Security] "...".
>
> Searching for my car in the vastly crowded parking lot, I scribbled down thoughts, affect, rage, questions about the (im)moral, (in)humane enactments of State sadism, the perverse layering of hell. I was ashamed that I didn't KNOW—shamed by my own epistemological ignorance of privilege, and the strategic recruitment of people of color into the cozy surveillance on people of color. Promising myself that I would find Tina, or investigate the flows of money and bodies, or at least tell you...

Conclusion

We have wandered through a series of borderlands where activist scholars, working in community, have taken up the project of decolonizing knowledges and carving spaces for liberatory inquiry in varied corners of the globe, at varied moments in history, within divergent movements and struggles (see, Adams et al., 2015). In the United States, it seems imperative to excavate a radical intellectual history of funky shoulders upon which we stand, and to bear witness today to slow violence as privilege accumulates, traces of fascism saturate, and new solidarities mobilize. Critical archives of research as resistance materialize then as minor deposits toward a full accounting of U.S. history; corrections to the record; counter narratives to provoke what Massey and Barreras (2013) call impact validity. Our field—critical community psychology—is well positioned to catalogue, with humility, the lives, communities and beliefs being slaughtered; the willful subjectivities, solidarities and forms of resistance being borne.

References

Adams, G., Dobles, I., Gomez, L., Kurtis, T., & Molina, L. (2015). Decolonizing psychological science: Introduction to the special thematic section. *Journal of Social and Political Psychology, 3*(1), 213–238.

Addams, J. (1910). *Charity and Social Justice: The Presidential Address at the National Conference of Charities and Correction* (pp. 441–449). Retrieved from https://digital.janeaddams.ramapo.edu/items/show/6294

Anzaldúa, G. (1987). *Borderlands/La frontera*. San Francisco: Aunt Lute.

Baldwin, J. (2017). *I am not your Negro*. New York: Penguin.

Bhatia, S. (2017). *Decolonizing psychology*. New York: Oxford University Press.

de Sousa Santos, B. (2014). *Epistemologies of the south: Against Epistemicide*. New York: Routledge.

Du Bois, W. E. B. (1899/1913). *The Philadelphia negro: A social study*. Philadelphia: Ginn.
Fine, M., & Ruglis, J. (2009). Circuits and consequences of dispossession: The racialized realignment of the public sphere for U.S. youth. *Transforming Anthropology, 17*(1), 20–33.
Foucault, M. (1972). *Archaeology of knowledge*. New York: Harper.
Gramsci, A. (1971). *Selections from prison notebooks*. New York: International Publishers.
Greenhaven Think Tank. (1997). *The non-traditional approach to criminal social justice*. New York: Center for NuLeadership, Resurrection Study Group.
Harding, S. (1994). Rethinking standpoint epistemology: What is 'strong objectivity'? In L. Alcoff & E. Potter (Eds.), *Feminist epistemologies* (pp. 49–82). New York: Routledge.
Human Rights First. (2018). *Ailing justice: New Jersey inadequate medical and mental health care in immigration detention centers*. New York: Human Rights First.
Jackson, J. P. (2004). 'Racially stuffed shirts and other enemies of mankind': Horace Mann Bond's parody of segregationist psychology in the 1950s. In A. S. Winston (Ed.), *Defining difference: Historical perspectives on psychology, race, and racism* (pp. 261–283). Washington, DC: American Psychological Association.
Johnson, G. (2018). *Guy Benton Johnson papers, 1830–1882, 1901–1987*. University of North Carolina, Louis Round Wilson Special Collections Library.
Lartly, J. (2017, December 28). *Private prison investors set for giant windfall*. Retrieved form www.theguardian.com/news/2017/dec/28/private-prisons
Lorde, A. (1984). *Sister outsider*. Berkeley, CA: Crossing Press.
Lykes, B. (Winter 1997). Activist participatory research among the Maya of Guatemala: Constructing meanings from situated knowledge. *Journal of Social Issues, 53*(4), 725–746.
Maldonado-Torres, N. (2016). *Outline of ten theses on coloniality and decoloniality. Fondation Frantz Fanon*. Retrieved from http://frantzfanonfoundation-fondationfrantzfanon.com/IMG/pdf/maldonado-torres_outline_of_ten_theses-10.23.16_pdf.
Martín-Baró, I. (1994). *Writings for a liberation psychology*. Cambridge, MA: Harvard University Press.
Massey, S., & Barreras, R. (2013). Impact validity as a framework for advocacy based research. *Journal of Social Issues, 69*(4), 615–632.
Montero, M., & Sonn, C. C. (2009). *Psychology of Liberation: Theory and applications*. New York: Routledge.
Narayan, U. (1998). Essence of culture and a sense of history: A feminist critique of cultural essentialism. *Hypatia, 13*(2), 86–106.
Patel, L. (2016). *Decolonizing educational research: From ownership to answerability*. New York: Routledge.
Payne, Y. (2013). The People's report: Co-Investigators: Charles A. Madden, J. D., Raye Avery Jones, Marlene Saunders, Ph. D. & Deborah Wilson, Hope Commission/University of Delaware, funded by First State Community Agency/American Recovery Reinvestment Act. Retrieved from www.thepeoplesreport.com
Rebaka, R. (2010). *Against epistemic Apartheid: WEB du Bois and the disciplinary decadence of sociology*. Lexington, MA: Lexington Books.
Sartre, J.P. (1961). *Preface to Fanon's Wretched of the Earth*. Retrieved from www.marxists.org/refereence/archive/sartre/1961
Segalo, P. (2011). Gendered suffering and the complexities of keeping silent. *Theological, Religious, and Ethical Voicelessness. In the New Millennium [Conference]*, Pretoria, South Africa.
Smith, L. T. (2005). On tricky grounds: Researching the native in an age of uncertainty. In N. Denzin & Y. Lincoln (Eds.), *The Sage handbook of qualitative research* (pp. 85–107). New York: Sage Publications.
Stevens, G., Duncan, N., & Hook, D. (2013). *Race, memory and the apartheid archive*. London: Palgrave MacMillan.

Stoudt, B. G., Torre, M. E., Bartley, P., Bracy, F., Caldwell, H., Downs, A., et al. (2015). "We come from a place of knowing": Experiences, challenges, advantages and possibilities of participating in Morris Justice Project. In C. Durose & L. Richardson (Eds.), *Re-thinking public policy making: Why coproduction matters*. Bristol, UK: Policy Press.

TheGuardian. (2017). *Private prison investors set for giant windfall from Trump tax bill*. Retrieved from https://www.theguardian.com/us-news/2017/dec/28/private-prisons-investors-trump-tax-bill

WITS Fallists. (2017). *Rioting and writing*. Retrieved from https://www.swop.org.za/single-post/2017/06/19/Rioting-and-writingwww.aclu.org/blog/immigrants-rights/immigrants-rights-and-detention

Wolgin, P. E. (2018). *Incarcerating entire families cannot be the solution to the separation of children*. Retrieved from https://www.americanprogress.org/issues/immigration/news/2018/06/20/452571/incarcerating-entire-families-cannot-solution-separation-children/.

Re/membering Ignacio Martín-Baró: Provocations and Insights Towards Liberating Psychology in the Twenty-First Century

M. Brinton Lykes and Gloria G. McGillen

> *Critical social psychology was not an abstract, intellectual exercise for Martín-Baró. Living within the 'limit situation' of the Salvadoran conflict, his choices were to be with 'them' or 'us.' He chose the side of the popular majority and he worked among the campesinos [peasants] in his many capacities as citizen, priest, educator, and social scientist. What made his engagement in the civil life of El Salvador a critical 'praxis,' however, was his persistent effort to understand the psychological dimensions of his experience and in so doing expand the scope of psychological theory and methods.... This essential connection between doing and thinking, commitment to the oppressed, and production of critical psychological insights creates the most challenging legacy of Martín-Baró's life and work.*
>
> —Liem, 1995, p. 443

The thirtieth anniversary of the assassination of social psychologist and Jesuit priest Ignacio Martín-Baró by members of the Salvadoran Army's Atlacatl Battalion—financed by the United States and trained at what was then the U.S. School of the Americas—is a stark reminder of the injustices of war and popular repression that were the focus of much of his scholarship at the time of his murder (Whitfield, 1994). Martín-Baró's life and works were bold responses to the conflicts of his time, shaped by and engaged with this socio-political context. The challenges his life project posed to psychological orthodoxy during the 1980s persist today and interface with a growing literature from decolonial and intersectional scholar activists of

M. B. Lykes (✉)
Boston College, Boston, MA, USA
e-mail: lykes@bc.edu

G. G. McGillen
University of Missouri, Columbia, MO, USA

G. Stevens, C. C. Sonn (eds.), *Decoloniality and Epistemic Justice in Contemporary Community Psychology*, Community Psychology,
https://doi.org/10.1007/978-3-030-72220-3_5

the Third and Fourth Worlds[1] as well as minoritized psychologists and practitioners in core capitalist countries. Many of them have taken inspiration from Martín-Baró while writing as members of groups marginalized from power who have accessed educational sites despite longstanding social inequities, impoverishment, and racism, as well as other global challenges emergent in the twenty-first century. His legacy extends through an increasingly international network of individuals and organizations who, we argue, seek to articulate a liberation psychology toward the new horizon Martín-Baró challenged psychologists to create. Yet, as we will also suggest, the discipline of psychology remains grounded in many of the same ontological and epistemological assumptions, limiting institutional practices, and problems of professional self-interest that he challenged in the 1980s. For example, and as will be discussed briefly below, the U.S.-based American Psychological Association (APA) persists in sustaining and benefiting from its support of the military-industrial defense system in which it successfully rooted its claims to being "a science" during and after WWII. This chapter presents some of the core challenges of Martín-Baró's liberation psychology of such practices, followed by a selective review of some of the ways in which psychologists have sought to re/member, elaborate, and extend them. We recognize that this is only a partial review and focus on those contributions that can help us answer the question—has Martín-Baró's work contributed to *liberating psychology* in the twenty-first century?

The concept of liberation evokes notions of overcoming oppression and securing a better future for those deprived of freedom. For Martín-Baró, liberating psychology in Latin America was a task whose urgency was provoked, at least in part, by the discipline's silence over El Salvador's repressive civil war, extreme poverty, and the dispossession of the majority population throughout Central America. He envisioned a future in which imperialistic foreign influence, government corruption and state violence, and a national academic apparatus obscuring these realities would give way to a more democratic society that would serve the needs of the majority of the Salvadoran people. Psychology would act as one tool for its realization, alongside and informed by the popular struggle and social movements of the majority. In the process, the discipline would lessen its deference to an oppressive social order and loosen itself from the stifling confines of the intellectual model of U.S. psychology.

As a Spanish-born Jesuit priest who alongside many of his fellow Jesuits took up a mission in El Salvador, Martín-Baró was a dedicated student of philosophy and liberation theology. Unlike most of his colleagues at the Universidad José Simeón Cañas (UCA) in San Salvador, however, he was also a social psychologist educated in the staunchly positivist U.S. tradition at the University of Chicago. He adopted

[1] There is no adequate label that captures the complex relationships among peoples of the globe. Some scholars resist language of First through Fourth Worlds, opting for global North or South, while others eschew these more recent geographically-based formulations. Despite their distinctions, each of these formulations re-inscribe hierarchies vis-à-vis a more powerful referent, most typically, Euro-American neoliberal North-based capitalism. We use the current discourse in solidarity with those seeking to critique the impoverishment and marginalization of their people who struggle towards a liberatory equity and justice.

some of the research methods from this training but also confronted epistemological, methodological, and ethical contradictions from which he generated his critique as well as his vision for a more emancipatory approach to psychology. He drew from liberation theology (Gutiérrez, 1988) a stance towards the poorest of the poor, that is, a "preferential option for the poor", as well as knowledge that social and political liberation require the elimination of the immediate as well as the root causes of poverty and injustice. He sought strategies through which he could build relationships that recognized and honored the dignity of the poor while facilitating processes through which they could share their knowledge with the wider Salvadoran community. Thus, despite the latter being accomplished through more conventional social psychological survey research strategies honed at the University of Chicago, Martín-Baró's community work and writing about his relationships with the Salvadoran people reflected the popular education methodologies of Paulo Freire (2000) and the community-based participatory research strategies of Orlando Fals-Borda (1987). He envisioned a Latin American psychology wherein the liberation of the individual person and the whole people (*el pueblo)* would be realized through "*a new praxis*." In calling for a "*new epistemology*" he emphasized the truth of the popular majority as constructed "from below", a process through which the oppressed would build their truth through action-reflection processes. The latter included recovering historical memory and de-ideologizing everyday experience (Martín-Baró, 1986, 1994).

Multiple U.S. community-based and research psychologists confronted local socio-political realities that challenged the assumptions of their discipline during the 1970s. Some formed part of one or more of the multiple social movements from the 1960s to 1980s, including civil rights protests, the second wave of the women's movement, the Black Power movement, and the Stonewall Uprising in New York City. The 1963 Community Mental Health Act and an anti-psychiatry movement, at the same time, deinstitutionalized mental health and laid the grounds for community mental health centers without adequate funding (Brown, 1973). These social movements contributed to some emerging critical ideas within the academy in the U.S. and beyond (see, for example, Gordon, 1973, on Black versus Euro-American psychology, and Zuñiga's discussion of the contributions of social psychology in times of radical social reform, 1975). As importantly, the latter challenged prevailing psychological thought as Martín-Baró began his graduate studies in social and organizational psychology at the University of Chicago in 1976.

Despite these incipient challenges from the margins, the mainstream intellectual culture of U.S. psychology in the 1960s and 1970s, and its growing international influence throughout the Cold War period, continued to be marked by positivism, methodologism, and ahistoricism—what Martín-Baró (1994) later described as "scientistic mimicry" (p. 20). As noted briefly above, psychology's alignment with state and corporate aims also strengthened in this period as psychologists continued to expand government and business relationships established throughout World Wars I and II (Danziger, 1994; Herman, 1995; Rose, 1998). Strikingly, as will be discussed below, these ties have persisted, as noted in the Hoffman Report (2015), which describes them as context for some APA members' and leaders' facilitation of the torture of individuals held in U.S. detention under the War on Terror. This modern example of the socially destructive potentials of psychology when the situations of the oppressed are left unacknowledged, and when theory and method contribute to obscuring them through neoliberal ideology and the self-interested pursuit of power, mirrors in important respects the conditions that led Martín-Baró (1994) to press for a "new horizon, new epistemology, and new praxis" (p. 26) for psychology in Latin America in the 1980s. The following selectively traces countercurrents to this dangerous status quo within the profession, from Martín-Baró forward through his legacy.

Many New Horizons: Contemporary Legacies of Martín-Baró

Throughout his academic career, Martín-Baró published frequently and, during the last decade of his life, spoke internationally about his vision for a liberation psychology. He developed relationships of solidarity with psychologists throughout the Americas and Europe in pursuit of visibility for these ideas and recognition of the urgent condition of Salvadorans (Dykstra, 2014; Gondra, 2013; Jiménez-Domínguez, 2009). His works have grown in circulation and influence in the 30 years since his death, aided by the dedication of colleagues and the contributions of his ideas for analyses of war and state violence, critiques of academic imperialism, and oppressed groups' development and emancipation within their gendered, political and socio-historical contexts. He has inspired intellectual and clinical innovations and academic training initiatives in both core capitalist countries and those on the global periphery (Burton & Kagan, 2009; Lykes & Moane, 2009; Montero, 2004, 2017). His work has also become a focus of dialogue in a number of scholarly and clinical networks—such as the Liberation Psychology Network and among participants in the International Conferences on Community Psychology—bringing together psychologists across countries and areas of specialty (Dykstra, 2014; González Rey, 2004; Sonn et al., 2017).

Through the lens of these contributions, we discuss how Martín-Baró's thought has been taken up, moving liberation psychology toward new horizons. We explore how a range of individuals and collectives have tried to operationalize what they have learned from Martín-Baró's *liberation psychology* as *psychologists*, frequently at odds with the discipline as typically practiced in their locales. We organize this discussion around three domains—scholarship and research, mental health practice, and academic and clinical training.

We use archival methods to identify some of the ways in which Martín-Baró's ideas have traveled through the printed word in these three areas from the latter years of his life to the present day. Despite enormous diversities in access to his work due to the majority of his publications appearing in Spanish, to English monolingualism prevalence among U.S. psychologists through the end of the twentieth century, and to the ongoing dominance of English in the psychological literature and attendant databases, our review finds a growing and diversifying body of transnational work that builds upon Martín-Baró's liberation psychology, defined by its pursuit of professional reflexivity and praxis, dedication to those who are marginalized and oppressed, and generativity of critical psychological insights (Liem, 1995). Despite the growing presence of his ideas within published psychological research and in a variety of graduate programs, the work reported continues to have a minor and marginalized presence within the overall field of psychology. That said, it has informed a developing body of interdisciplinary and intersectional theory and research from activist scholars, many of whom are from the Third and Fourth World communities whom Martín-Baró sought to accompany during the Salvadoran armed conflict as well as from minoritized groups within core capitalist countries. We report on some of the ways in which his ideas are being transformed by Indigenous, feminist, anti-racist scholars, among others, and their challenges in the twenty-first century.

Scholarship and Research

During his career, Martín-Baró displayed an overriding concern with the political implications of academic work for both oppressed and powerful groups, including psychological professionals. His methods as a scholar and empirical researcher were rigorous and diverse but deeply threatened not only those who dominated the profession at the time but also the Salvadoran government and military. He often spoke—seemingly in jest but poignantly presciently—of the pressures on U.S. psychologists to "publish or perish", while in El Salvador, those who published risked perishing. By the time of his murder, he had established himself as an internationally recognized thinker and researcher and an influential figure in psychology in Latin America (Gondra, 2013; González Rey, 2004; Marín, 1991).

Between 1966 and November 1989, Martín-Baró produced a body of scholarly work that included approximately 113 articles, books, speeches, and other media, the vast majority in Spanish (Martín-Baró, 1994). Several additional works appeared posthumously through the efforts of colleagues. Gerardo Pacheco and Bernardo Jiménez (1990) published the first anthology of Martín-Baró's writings outside of El Salvador through the University of Guadalajara in Mexico. English-language translations of Martín-Baró's writings began to appear shortly thereafter, beginning with several articles in *Towards a Society That Serves Its People: The Intellectual Contribution of El Salvador's Murdered Jesuits* (Hassett & Lacey, 1991) and followed by *Writings for a Liberation Psychology* (Martín-Baró, 1994), edited by Adrianne Aron and Shawn Corne.

Theoretical and Methodological Contributions towards Liberating Psychology. The tenor of Martín-Baró's writings about and leadership within organized psychology was decidedly critical. Drawing on his 1986 publication in the *Boletín de Psicología* and subsequent lectures in a variety of psychological conferences in 1987 and 1989 (see Montero, 2004), he described the new horizon, epistemology, and praxis that he envisioned for a Latin American social psychology of liberation.

He challenged Latin American psychologists to focus on the needs of the "majority population" rather than on the discipline's scientific and social status, situating the liberation of the individual person within the context of *el pueblo*, the whole people or community. Using the examples of his survey research through which he identified the opinions of Salvadoran rural peasants whose ideas about the armed conflict had not been represented in the mainstream press or taken up in public policies, he argued that the truth of the popular majority must be constructed "from below." These ideas echoed the aforementioned liberation theology's "preferential option for the poor."

Another challenge was methodological and drew on both Fals-Borda and Freire, emphasizing psychology's need for a new praxis, that is, for engaging with community groups or organizations in action-reflection processes. This new psychology would be grounded in the people's reality and foster the liberation of all people—the community and the psychologist, as well as the latter's discipline. This praxis utilized people's virtues in recovering historical memory and deideologizing everyday experiences (Martín-Baró, 1986, 1994). Gondra (2013) described the latter as a process through which one retrieves.

> the original experience of groups and persons in order to articulate a consciousness of their own reality, and by so doing, verify the validity of acquired knowledge… [a process] realized through critical participation in the life of the poor, which represented a certain departure from the predominant forms of research (p. 14).

Initial Contributions to a Liberation Psychology. As a journal editor, Martín-Baró sought to cultivate international dialogues about these ideas and offered a platform through the *Revista de Psicología de El Salvador* to colleagues like Bernardo Jiménez-Domínguez of Mexico (1988), Maritza Montero of Venezuela (1988), and Ramsay Liem (1988) and Tod Sloan (1988) of the U.S. who represented the critical edge of the discipline in their respective countries. Although his assassination alongside UCA colleagues Ignacio Ellacuría, Amando López, Joaquín López y López, Segundo Montes, and Juan Ramón Moreno, the caretaker of the rectory at UCA, Elba Julia Ramos, and her daughter, Celina Ramos, drew international condemnation of the conditions of Salvadorans and the role of U.S. and international policy in sustaining the country's armed conflict (Lassalle-Klein, 2014; Leogrande, 1990; Whitfield, 1994), the first decade after his murder saw relatively few publications in any language that drew on his nascent liberation psychology (Montero, 2004). Two articles referencing Martín-Baró's work had appeared in journals indexed internationally through Scopus during the 1980s (i.e., Montero & Sloan, 1988; Montero, 1989). These references increased to only 44 in the subsequent decade (see Table 1). Despite that, some psychologists were beginning to respond to his call and that of liberation theologians to ground their community-based work in the needs and perspectives of those experiencing oppression and marginalization against the backdrop of ongoing war and deadly violence (see, also, Montero, 2004). This work would slowly take root among Latin American and some North American and European psychologists and be articulated in their publications. Early articles explored the situation of Salvadoran refugees from the civil war (e.g., Farias, 1991;

Table 1 Citations of Martín-Baró's works in peer-reviewed psychology publications indexed in Scopus, 1980–2019

Decade	No. of articles	Author countries	Author regions	Top five ranked journals by productivity
'80–'89	2	3	Latin America: 2 U.S. & Canada: 1	*International Journal of Psychology* (1) *Revista de Psicologia Social* (1)
'90–'99	44	9	U.S. & Canada: 28 Latin America: 8 Europe: 7	*Social Science and Medicine* (8) *American Psychologist* (4) *Revista Panamericana de Salud Publica* (3) *American Sociologist* (2) *Culture, Medicine, and Psychiatry* (2) *Journal of Community Psychology* (2) *Journal of Social Issues* (2) *Political Psychology* (2) *Psychology of Women Quarterly* (2) *Revista de Psicologia Social* (2) *Theory & Psychology* (2) *Women and Therapy* (2)
'00–'09	216	29	U.S. & Canada: 107 Latin America: 59 Europe: 52 Australasia: 4 Africa: 3 Asia: 3 Middle East: 2	*The Counseling Psychologist* (15) *American Journal of Community Psychology* (12) *Feminism and Psychology* (12) *Interamerican Journal of Psychology* (12) *Journal of Community and Applied Social Psychology* (11) *Universitas Psychologica* (10) *Journal of Health Psychology* (7) *Psicologia E Sociedade* (7)
'10–'19	409	39	U.S. & Canada: 183 Latin America: 150 Europe: 113 Australasia: 17 Africa: 13 Middle East: 4 Asia: 3 Caribbean: 1	*Psicologia E Sociedade* (22) *Peace and Conflict* (21) *American Journal of Community Psychology* (18) *Journal of Prevention and Intervention in the Community* (15) *Universitas Psychologica* (15) *Journal of Social and Political Psychology* (13)

Jenkins, 1991) as well as the traumatic effects of the armed conflict on children in rural Guatemala (e.g., Lykes, 1994). Others expanded their attention to issues of ethnic conflict, state violence, post-conflict situations, and refugee movements in countries beyond Central America, including Peru, Puerto Rico (Comas-Díaz, Lykes, & Alarcón, 1998), Bosnia (Jones, 1998), and the U.S. (Adams, Barton, Mitchell, Moore, & Einagel, 1998). A final area addressed was the value of community-based and participatory action research as socially just methods of investigation with the potential to promote constructive social change (e.g., Lykes, 1997; Sánchez & Almeida, 1992, among others in Mexico).

Scope of his Intellectual Influence. U.S. and Canadian psychologists contributed many of the articles drawing upon Martín-Baró's thought in this decade indexed in Scopus (Table 1). Additional but substantially smaller centers of scholarly activity appeared in Central and South America and Europe—led by José Miguel Cruz in El Salvador (1999) and Ignacio Dobles in Costa Rica (1999). This period of rapid growth in English-language scholarship was likely a result, in part, of the English translations of Martín-Baró's works newly available by the mid-1990s and the solidarious relationships he had established while urging North American psychologists to transform their profession (Dykstra, 2014). Yet the continuing hegemony of U.S. and other western institutions in the discipline, favoring the visibility on the international stage of scholars situated in these institutions and the positivist methods they demand (Fisher, Sonn, & Evans, 2007; Guzzo, 2015) is reflected in the publications themselves, some of which continue to employ traditional research frameworks and include limited discussion of the sociopolitical and historical context of their topics of inquiry. This pattern persists today and is particularly evident in publications focused domestically on conditions in North America, despite the extensive histories of violence, oppression, and popular resistance on the continent (Burton & Kagan, 2009; Jemal, 2017).

The circulation of Martín-Baró's thought continued to grow in the first two decades of the twenty-first century, despite these challenges, generating a total of 625 referenced publications from 2000 to mid-2019. In contrast with the 1990s, these decades saw greater parity in publication of internationally-indexed articles from scholars based in Latin America, many writing in Spanish and Portuguese, and those in the U.S. and Canada. The top-producing journal of the decade by 2019, for example, was the Brazilian Association of Social Psychology's (ABRAPSO) *Psicologia E Sociedade*, which overtook more U.S.-based publications such as *The Counseling Psychologist* and the *American Journal of Community Psychology* from the prior decade, driven in large part by the work of Brazilian liberation social psychologists such as Raquel Guzzo (2015). Evident in these decades is also the growing influence and international visibility of Martín-Baró's thought in the work of authors from Africa, East and Southeast Asia, the Middle East, and Australia and New Zealand, including, among others, Ibrahim Makkawi (2015, 2017), Christopher Sonn and colleagues (2017), Simone Lindorfer (2007), Cristina Montiel and colleagues (2009), Chaudhry & Bertram (2009), Desmond Painter and Martin Terre Blanche (Painter & Terre Blanche, 2004), and Derek Hook (2005) who have established generative lines of research and scholarship informed by Martín-Baró's ideas.

Despite the interest in a more liberatory psychology evidenced in these international contexts and ensuing publications, psychology in the U.S. has remained engaged in important respects in the academic imperialism that characterized the field when Martín-Baró was assassinated. The direct collaboration of research and applied psychologists in facilitating torture during U.S. government interrogations abroad, and the slow and uneven professional response within APA membership to the earliest warnings about these violations, provided jarring evidence of this reality (Teo, 2015). These disturbing breaches of psychological ethics and the public trust were publicly confronted by those outside the discipline (McCoy, 2006) as well as by a small *Coalition for an Ethical Psychology* including, among others, Steven Reisner, Roy Eidelson, Stephen Soldz, and Jean Maria Arrigo who detailed how APA leaders enabled U.S. psychologists to participate in military and C.I.A. torture of individuals detained at sites such as Guantanamo Bay. These efforts contributed to APA's appointment of independent investigators with the law firm Sidley Austin, LLP and the publication by APA of the Hoffman Report (2015) which, concluded, among other things, that,

> The end result of… the [APA] Ethics Office's purposeful unwillingness to thoroughly investigate allegations of unethical conduct by psychologists who participated in interrogations was that the Ethics Office prioritized the protection of psychologists—even those who might have engaged in unethical behavior—above the protection of the public (p. 63).

Such behaviors, and institutional failures to swiftly and critically expose and condemn them, demonstrate the challenge that psychologists' pursuit of professional interests above the needs of the wider society has continued to pose since Martín-Baró's critique of the profession of psychology. It further speaks to the urgency of his call to liberate psychology from itself and the barriers those who have embraced his vision for the profession face in influencing its dominant practice. Prior to exposés in the U.S. press, psychologists across several countries (e.g., Burton & Kagan, 2007; Costanzo, Gerrity, & Lykes, 2006; Lira, 2008) drew upon Martín-Baró's example to confront the profession's involvement in torture and other forms of state violence under the global War on Terror. More recent work (e.g., Eidelson, 2014, 2019) has also addressed evidence of psychologists' collaboration in projects promoting the use of weaponized drones in warfare, drawing attention to findings that the use of military drones "terrorizes men, women, and children, giving rise to anxiety and psychological trauma among civilian communities" (Cavallaro, Knuckey, & Sonnenberg, 2012, p.vii).

Other work on psychologists' responsibility to address state violence has considered political conditions in Zimbabwe (Mpande et al., 2013), Palestine (Makkawi, 2015, 2017; Sousa & Marshall, 2017), Mexico (Sánchez & Trujano Ruíz, 2011), Chile (Cornejo, Morales, Kovalskys, & Sharim, 2013; Sharim, Kovalskys, Morales, & Cornejo, 2011), and Argentina (Arnoso Martinez, Arnoso Martinez, & Pérez Sales, 2015). Despite this urgently needed work, often undertaken at professional and personal risk to these psychologists and their collaborators, it remains outside the professional mainstream.

Some contemporary psychologists' express skepticism that individuals acting in their roles as psychological professionals can pursue justice and challenge power effectively, given entrenched epistemologies and practices within the disciple and its dedication to self-interest (e.g., Fox, 2008; Walsh & Gokani, 2014). This as well as the tendency of psychologists to psychologize the political, supports their claim that much, if not most, social change will depend on externally grounded popular movements. To wit, the 30 years since Martín-Baró's assassination have seen the continued evolution of diverse social movements of communities historically marginalized from access to power and resources (Tilly, Castañeda, & Wood, 2019). Yet, in contrast with the skeptical vision of a separation between the profession and social movements, some psychologists have drawn from them inspiration and motivation for collaborating in liberatory, feminist, and anti-racist praxis (e.g., Dutta, Sonn, & Lykes, 2016; Fine, 2012, 2013; French et al., 2019; Lykes & Moane, 2009; Watts, 2004; Watts, Griffith, & Abdul-Adil, 1999). Others have documented multiple and diverse ways of knowing from within their local and Native communities (Duran, Firehammer, & Gonzalez, 2008; Gone, 2008, 2013), challenging the discipline from within. They have pressed for a decolonization not only of psychological research but also of its teaching and praxis (see, e.g., Ronelle L. Carolissen and Paul Duckett's two volume Special Issue of the *American Journal of Community Psychology*, 2018, volume 62, issues 3–4). Some of these activist scholars draw on Martín-Baró's ideas whereas others describe themselves as, for example, social community or critical social psychologists (Burton & Gómez Ordóñez, 2015). Martiza Montero (2004) has developed an illuminating comparison among these multiple iterations of what the mainstream or white-stream discipline might call "progressive psychologies". She suggests that community and liberation psychologists tend, for example, to draw on Freirian concepts (e.g., problematization or conscientization) while critical psychologists are more likely to critique core concepts in psychology, for example, the individual subject vis-à-vis multiple subjectivities or positionalities (see, Gergen, 1991; Henriques, Hollway, Unwin, Venn, & Walkerdine, 1984). Community and critical psychologies tend towards more participatory and activist research strategies with the former prioritizing the meaning-making of community participants and co-researchers while critical psychology shares their critiques of positivism while drawing more on phenomenological research strategies (Grabe & Dutt, 2015). Despite their differences, Montero (2004) emphasizes the convergence of many of these ideas and argues that taken together they may consolidate an alternative praxis that can contribute to the "transformation of these societies … towards a more ecologically viable and humanly desirable development" (p. 27). As suggested below, more recent activist and minoritized scholars (e.g., Bell, Canham, Dutta, & Fernández, 2019) draw on Martín-Baró's ideas while pressing for decolonizing the discipline and beyond.

Mental Health Practice

While Martín-Baró was not trained in a clinical discipline within psychology, his vocation as a Jesuit priest and pastoral relationships with Salvadorans across decades brought him into daily contact with human distress and challenged him to articulate his work in ways that fostered healing within the context of armed conflict and societal repression (Lindorfer, 2007; Martín-Baró, 1994). He wrote and lectured on clinically significant topics crafting the language of "psychosocial trauma" to capture the individual and social phenomena he observed in the ongoing conflict in El Salvador. He documented the psychological development of peasants and youth in situations of historical oppression and described collective, community-based healing practices, including social movement participation. These contributions have remained prominent in some clinical writings that draw upon Martín-Baró's work. Additionally, new concerns, including the psychological needs of diverse populations of the world's growing number of migrants and refugees and of historically oppressed groups in core capitalist countries such as the U.S., have also taken hold within this literature (e.g., Afuape, 2012; French et al., 2019; Hernández-Wolfe, 2011, 2013; Singh, 2016; Watkins & Shulman, 2008).

Consistent with other early publications drawing upon Martín-Baró, many of the first writings using his ideas focus on the psychological needs of survivors of Central America's armed conflicts, particularly children (e.g., Boothby, 1992; Melville & Lykes, 1992; Quesada, 1998; Walton, Nuttall & Nuttall, 1997). Many challenged individualist models of psychological distress and healing, following Martín-Baró's liberationist model. Bracken, Giller, and Summerfield (1995), for example, summarizing the conflict in El Salvador, argued that "what was left traumatised were not just Salvadoran individuals, but Salvadoran society" (p. 10). Lykes (1997) argued that a western psychological focus on trauma as an individual-level rupture in describing Mayan survivors ignored their narratives, stories that reflect Indigenous sociality and community. The Maya articulation of self, of "who I am", invokes family, community, the animal kingdom, traditions and language, and the earth. Fischman (1998) offered a wide-ranging discussion of the societal implications of traumas inflicted upon individuals, including torture and other acts of direct and systemic violence. Comas-Díaz et al. (1998) drew on and extended Martín-Baró's language of the psychosocial, discussing the functions of torture and terror in the context of genocidal violence and armed conflict through a focus on their symbolic dimensions. They suggested that the latter extends both the spatial and temporal dimensions along which terror and trauma was more typically conceptualized, arguing that its destructive forces reach beyond the individual to both the community and the cultural context, noting that,

> Attention to the symbolic dimensions of terror and to the symbolic representations of terror's effects is, therefore, critical for approximating knowledge of Mayan stories of victimization and survival and furthers psychology's theories about trauma, its consequences, and the healing process (Comas-Díaz et al., 1998, p. 783).

Anckermann and colleagues (2005) have more recently presented a three-point guide for understanding and intervening in the wake of psychosocial trauma, developed for use in large-scale community interventions in post-conflict settings in Central America.

Innovative approaches to emancipation and healing suggested by Martín-Baró, such as problemtaization, conscientization, and deideologization, also appear in clinicians' theorizing across decades. Lindorfer (2007) drew on the latter construct in her analysis of psychosocial trauma in Central Africa, focusing upon the need for mental health professionals to carry on this process personally and within their organizations as well as on the limits of liberation. She writes:

> De-ideologizing, as a professional individual and organizational attitude, means acknowledging that there is no innocent standpoint in helping. It calls for a constant re-viewing of ourselves, our tools and methods, our implicit ideologies and our legitimizing role of international power structures (pp. 364–365).

Moane (2003) presented a workshop model focused on the social analysis or deideologization of homophobia to promote wellbeing and camaraderie among intergenerational groups of Irish lesbians, and Singh (2016) enumerated multiple strategies for emancipatory work with people who are transgender and gender nonconforming (TGNC), emphasizing therapists' responsibility to "deideologize psychology for TGNC clients by explicitly naming the oppressive history within mental health care provision of services over the last few decades and intentionally exploring client issues of distrust" (p. 758). French et al. (2019) explored conscientization as a key element in their paradigm for community-based radical healing for people of color and Indigenous people (POCI) in the U.S., a process through which they suggested allied psychologists should "engage in direct actions of social advocacy and political resistance to disrupt the oppressive systems that plague our society and prevent POCI's liberation" (p. 23). Watkins and Shulman (2008) critically analyze social suffering in unjust and violent contexts in a wide variety of global communities towards identifying local strategies from various parts of the world wherein survivors develop shared strategies including participatory and action research, the creative arts and other liberatory praxis towards renarrating their individual and collective lives towards healing.

Current levels of displacement and migration of vulnerable groups worldwide, which analyses suggest may be at their highest levels since World War II (Committee on Foreign Relations, U.S. Senate, 2020; United Nations High Commission for Refugees, 2020), present complex psychosocial challenges that a growing number of authors have also taken up through the framework of liberation psychology. Afuape (2012), and many contributors to the collection edited by Afuape and Hughes (2016), have examined issues of clinical practice with migrants to metropolitan areas in the U.K., particularly youth and young adults navigating bicultural development, racial oppression, and the effects migration-related stress and trauma. They presented methods of healing that include elements of narrative therapy, media and textual analysis, and social activism intended to facilitate youth's movement through a cycle of critical reflection and therapeutic action in society. They also highlighted systemic barriers within the U.K. National Health Service's (NHS) dominant medical model of care that constrain clinicians and clients in their pursuit of these practices. Cautiously yet optimistically, however, and following the words of one client, they conclude about these constraints, "Even when you feel like you're trapped in a box, you can still make holes in it" (p. 210).

Drawing upon their extensive work with North African migrants to Spain, Paloma, García-Ramírez, Camacho, and Olmedo (2016), finally, have discussed potentials for migrants to be supported in transforming structural conditions that they encounter throughout their processes of migratation and adaptation to a new society and culture. Their body of work challenges psychologists working in the liberation tradition to confront organizational barriers such as those detailed by Afuape and Hughes within the NHS by applying its insights to organizational development and management. This call highlights the still limited but growing movement in contemporary clinical adaptations of Martín-Baró's ideas to reengage the project of structural analysis and systems change within health care that marked activist efforts in the 1960s and 1970s in the U.S. leading to the 1963 Community Mental Health Act, as well as similar past and ongoing social movements in countries throughout Latin America, Africa, and elsewhere to transform dominant systems of mental health care and psychological healing (see, e.g., Painter & Blanche, 2004; Smith-Nonini, 2010; Waitzkin, Iriart, Estrada, & Lamadrid, 2001; Yen, 2016).

Academic and Clinical Training

Underlying Martín-Baró's scholarly accomplishments and social impact, and those of the individuals who have been influenced by his work, are the institutional contexts they inhabit—largely, international universities as faculty, researchers, and students. Martín-Baró and his colleague Ignacio Ellacuría encouraged their audiences to analyze university life as socially and politically embedded and to act upon it with an intention to improve their societies. Ellacuría (1991), reflecting on this issue in the essay "Is a Different Kind of University Possible?", summarized:

> To a great extent the university responds in each case to the greatest social pressure exerted on it and/or pressure from the state, and consequently it is shaped politically by these pressures, while at the same time and with a variety of direct and indirect means it can exert pressure on the power of society and the state. In itself it is a power to be reckoned with, especially if it could be given the cohesion it should have (p. 178).

Martín-Baró, Ellacuría, and their Jesuit colleagues' leadership of the UCA from the 1960s through 1980s, informed by the teachings of liberation theology, followed the belief that a Christian university must act concretely for popular freedom. They advanced this cause through scholarship, public opinion research, educating new generations of scholars and professionals, and public and political engagement within and beyond El Salvador (Ellacuría, 1991; Martín-Baró, 1994). Their actions reflecting these ideals caused them to be politically targeted, and eventually murdered, by Salvadoran armed forces, a testimony to the threatening social power they had accumulated in the eyes of the state by that time.

Their ideas about the social power of universities have served as a touchstone within and across many countries since their deaths, particularly vis-à-vis universities' potential and responsibility to train psychologists in a manner that contributes to their development of critical consciousness and a commitment to praxis (e.g., Cordero, 2003; Fisher et al., 2007; Goodman et al., 2004; Robledo Gómez, 2008; Singh et al., 2010; Speight & Vera, 2004; Wallace, 2000; Watts, 2004). Academic and clinical training programs and seminar series built upon these goals continue in the UCA in El Salvador (Portillo, 2012) and have emerged in multiple other locations in Latin America, including, for example, annual meetings of the Catédra Libre Ignacio Martín-Baró in Colombia (https://psicologia.javeriana.edu.co/catedra-martin-baro) and the Catédra Martín-Baró in Chile (https://psicologia.uahurtado.cl/xiii-catedra-ignacio-martin-baro-2020/). Geraldine Moane's classes at the University College Dublin reflect a liberationist approach to psychology

(Dykstra, 2014) whereas, in the U.S., Pacifica Graduate Institute's master's and Ph.D. programs in Community, Liberation, Indigenous, and Eco-Psychologies, established by Mary Watkins among others, are probably the most systematic curricula reflecting Martín-Baró's liberation psychology while also incorporating more recent developments from Indigenous and ecological psychologies. Another initiative within the U.S. that has sought to sustain the vision and priorities of liberation psychology as exercised through community groups of the majority world is the Ignacio Martín-Baró Fund for Mental Health and Human Rights (Lykes, 2012; Lykes & Sibley, 2014; www.martinbarofund.org). Since its founding in 1989, the fund has raised and distributed over $1.5 million through volunteer efforts to support grassroots initiatives of marginalized groups seeking to redress harm from, for example, gendered and racialized violence and armed conflict, while organizing as local survivors to engage transformative change strategies that draw on their traditional knowledge to ensure their personal and collective well-being. In 2020 it extended this work and is now the Martín-Baró Initiative for Wellbeing and Human Rights at Grassroots International (https://grassrootsonline.org/martin-baro-initiative-for-wellbeing-and-human-rights/).

The accomplishments of these initiatives and the barriers they face are influenced by institutional and national conditions as well as, increasingly, global circulations of power that constrain or facilitate scholars' ability to conduct and disseminate work challenging racism, imperialism, settler colonialism, and patriarchy, their interrelations, and their academic manifestations. This is especially true of scholars in the Third and Fourth Worlds, as is vividly captured in Makkawi's (2017) work to chronicle the "rise and fall" of the academic community psychology program he developed at Birzeit University in Palestine. Under the conditions of Israeli occupation of Palestine, he notes, psychologists face a "colonial context [where] economic dependency on Western funders is most likely to result in the academic dependency of the colonised" (p. 490).

The challenges posed by racism, neocolonialism, and imperialist models of academic life under conditions of greater economic "freedom" are also explored in Yen's (2016) account of attempts to develop a truly liberatory model of health psychology in apartheid and post-apartheid South Africa. Bell et al.'s (2019) retrospective autoethnographies as immigrant psychologists of color, a majority women, situated in and working to transform historically white universities in the U.K., U.S., and South Africa powerfully summarize the intentions of these and other members of a growing collective of psychologists worldwide who are carrying on the work of liberating the disciple of psychology and those it is intended to serve:

> [We] choose to build a university that has yet to be invented.... [R]adical movements of consciousness, born of the imagination and our imperatives, are forces, readily available to us, for building a university we desire. The struggles we name are systemic and yet the resultant suffering is seen as a form of private weakness, a mark of personal failure. Resisting and surviving an academy that finds new ways to inform us we do not belong is depleting. But for many of us, we are not in the academy because of an abstract love for knowledge or institutions. It is a political and ethical imperative, powerfully articulated by

liberation psychologist Martín-Baró (1994, pp. 41–42): to "make a contribution toward changing all the conditions that dehumanize the majority of the population, alienating their consciousness and blocking the development/ of their historical identity" (p. 16).

Concluding Thoughts and Challenges

We argue here that despite the growing influence of Ignacio Martín-Baró's life and work, challenges posed by psychological orthodoxy during the 1980s persist today. Despite the important contributions of Indigenous, feminist, anti-racist, and/or critical scholars and practitioners who have taken up Martín-Baró's ideas and write from and engage psychosocial praxis "against the grain", they remain at the margins of the discipline.

This chapter discusses the work of some within an international network of individuals and organizations who strive to liberate psychology toward the new horizon Martín-Baró challenged psychologists to create. This selective review discusses how some from related or distant sub-disciplines within psychology have sought to elaborate on Martín-Baró's theory and praxis, extending the scope of his methodologies to ground their work not only in participatory and action research but in the traditional knowledge(s) and wisdom of their communities. These twentieth and twenty-first century Indigenous, feminist, critical and anti-racist activist scholars extend Martín-Baró's liberation psychology forward and backward in time and space, demonstrating its potential as a praxis of healing that simultaneously challenges those who engage it to *liberate psychology* in the twenty-first century.

Yet, as we also suggest, the discipline of psychology remains grounded in many of the same assumptions and professional practices that Martín-Baró critically engaged and sought to change in the 1980s. As suggested above, psychologists are challenged to both liberate as well as to decolonize not only the dominant theories and research methods of the discipline of psychology but also the material practices and power relations in which our teaching, our training, and our community work are embedded. The activist psychologists' praxis discussed herein offers one small contribution towards sustaining and extending Martín-Baró's legacy in the twenty-first century.

References

Adams, D. P., Barton, C., Mitchell, G. L., Moore, A. L., & Einagel, V. (1998). Hearts and minds: Suicide among United States combat troops in Vietnam, 1957–1973. *Social Science & Medicine, 47*(11), 1687–1694.

Afuape, T. (2012). *Power, resistance and liberation in therapy with survivors of trauma: To have our hearts broken*. London: Routledge.

Afuape, T., & Hughes, G. (2016). *Liberation practices towards emotional wellbeing through dialogue*. London: Routledge.

Anckermann, S., Dominguez, M., Soto, N., Kjaerulf, F., Berliner, P., & Naima Mikkelsen, E. (2005). Psycho-social support to large numbers of traumatized people in post-conflict societies: An approach to community development in Guatemala. *Journal of Community & Applied Social Psychology, 15*(2), 136–152.
Arnoso Martinez, M., Arnoso Martinez, A., & Pérez Sales, P. (2015). Argentina (1976-1983): Psychosocial consequences and coping. *Universitas Psychologica, 14*(3), 833–842.
Bell, D., Canham, H., Dutta, U., & Fernández, J. S. (2019). Retrospective autoethnographies: A call for decolonial imaginings for the new university. *Qualitative Inquiry*. https://doi.org/10.1177/1077800419857743
Boothby, N. (1992). Displaced children: Psychological theory and practice from the field. *Journal of Refugee Studies, 5*(2), 106–122
Bracken, P. J., Geller, J. E. & Summerfield, D. (1995). Psychological responses to war and atrocity: The limitations of current concepts. *Social Science and Medicine, 40*(8), 1073–1082.
Brown, P. (1973). *Radical psychology*. London: Tavistock.
Burton, M., & Gómez Ordóñez, L. H. (2015). Liberation psychology: Another kind of critical psychology. In I. Parker (Ed.), *Handbook of critical psychology* (pp. 248–255). London: Routledge.
Burton, M., & Kagan, C. (2007). Psychologists and torture: More than a question of interrogation. *The Psychologist, 20*(8), 484–487.
Burton, M., & Kagan, C. (2009). Towards a really social psychology: Liberation psychology beyond Latin America. In M. Montero & C. Sonn (Eds.), *Psychology of liberation: Theory and applications* (pp. 51–72). New York: Springer.
Cavallaro, J., Knuckey, S., & Sonnenberg, S. (2012). *Living under drones: Death, injury, and trauma to civilians from US drone practices in Pakistan.* Standford Law School International Human Rights and Conflict Resolution Clinic and New York University Law School Global Justice Clinic. Retrieved from https://www-cdn.law.stanford.edu/wp-content/uploads/2015/07/Stanford-NYU-Living-Under-Drones.pdf
Chaudhry, L. N., & Bertram, C. (2009). Narrating trauma and reconstruction in post-conflict Karachi: Feminist liberation psychology and the contours of agency in the margins. *Feminism & Psychology, 19*(3), 298–312.
Comas-Díaz, L., Lykes, M. B., & Alarcón, R. D. (1998). Ethnic conflict and the psychology of liberation in Guatemala, Peru, and Puerto Rico. *American Psychologist, 53*(7), 778–792.
Committee on Foreign Relations, U.S. Senate, Minority Staff Report. (2020, June 18). *Global forced migration: The political crisis of our time* (Senate Prints 116-48, 116th Congress, 2nd session). https://www.govinfo.gov/content/pkg/CPRT-116SPRT41888/html/CPRT-116SPRT41888.htm
Cordero, T. C. (2003). La psicología en Costa Rica: Encrucijada en el nuevo milenio [Psychology in Costa Rica: Crossroads in the new millennium]. *Interamerican Journal of Psychology, 37*(1), 93–108.
Cornejo, M., Morales, G., Kovalskys, J., & Sharim, D. (2013). La escucha de la tortura desde el Estado: la experiencia de los profesionales de la Comisión Nacional sobre Prisión Política y Tortura en Chile [Listening to torture from the state: The experience of professionals in the National Commission of Political Prison & Torture in Chile]. *Universitas Psychologica, 12*(1), 271–284.
Costanzo, M., Gerrity, E., & Lykes, M. B. (2006). Psychologists and the use of torture in interrogations. *Analyses of Social Issues and Public Policy (ASAP), 6*(1), 1–14.
Cruz, J. M. (1999). El impacto psicosocial de la violencia en San Salvador [The psychosocial impact of violence in San Salvador]. *Revista Panamericana de Salud Pública, 5*(4/5), 295–302.
Danziger, K. (1994). *Constructing the subject: Historical origins of psychological research.* Cambridge, UK: Cambridge University Press.
Dobles, I. (1999). Marxism, ideology and psychology. *Theory & Psychology, 9*(3), 407–410.
Duran, E., Firehammer, J., & Gonzalez, J. (2008). Liberation psychology as the path toward healing cultural soul wounds. *Journal of Counseling & Development, 86*(3), 288–295.

Dutta, U., Sonn, C. C., & Lykes, M. B. (2016). Situating and contesting structural violence in community-based research and action. *Community Psychology in Global Perspective, 2*(2), 1–20.
Dykstra, W. (2014). Liberation psychology – A history for the future. *The Psychologist, 27*(11), 888–891.
Eidelson, R. (2014, 12 August). Predators, reapers, and psychology's do-no-harm ethics. *Psychology Today.* Retrieved from https://www.psychologytoday.com/nz/blog/dangerous-ideas/201408/predators-reapers-and-psychology-s-do-no-harm-ethics?amp
Eidelson, R. (2019, 22 April). War, militarism, and the APA. *Psychology Today.* Retrieved from https://www.psychologytoday.com/us/blog/dangerous-ideas/201904/war-militarism-and-the-apa
Ellacuría, I. (1991). Is a different kind of university possible? In J. J. Hassett & H. Lacey (Eds.), *Towards a society that serves its people: The intellectual contribution of El Salvador's murdered Jesuits* (pp. 177–208). Washington, DC: Georgetown University Press.
Fals-Borda, O. (1987). The application of participatory action-research in Latin America. *International Sociology, 2*(4), 329–327. https://doi.org/10.1177/026858098700200401
Farias, P. J. (1991). Emotional distress and its socio-political correlated in Salvadoran refugees: Analysis of a clinical sample. *Culture Medicine and Psychiatry, 15*, 167–192. https://doi.org/10.1007/BF00119043
Fine, M. (2012). Troubling calls for evidence: A critical race, class and gender analysis of whose evidence counts. *Feminism & Psychology, 22*(1), 3–19. https://doi.org/10.1177/0959353511435475
Fine, M. (2013). Echoes of Bedford. A 20-year social psychology memoir on participatory action research hatched behind bars. *American Psychologist, 68*(8), 687–698.
Fischman, Y. (1998). Metaclinical issues in the treatment of psychopolitical trauma. *American Journal of Orthopsychiatry, 68*(1), 27–38.
Fisher, A. T., Sonn, C. C., & Evans, S. D. (2007). The place and function of power in community psychology: Philosophical and practical issues. *Journal of Community & Applied Social Psychology, 17*(4), 258–267.
Fox, D. (2008). Confronting psychology's power. *Journal of Community Psychology, 36*(2), 232–237.
Freire, P. (2000). *Pedagogy of the oppressed* (30th anniversary ed.). New York: Continuum.
French, B. H., Lewis, J. A., Mosley, D. V., Adames, H. Y., Chavez-Dueñas, N. Y., Chen, G. A., et al. (2019). Toward a psychological framework of radical healing in communities of color. *The Counseling Psychologist.* https://doi.org/10.1177/0011000019843506
Gergen, K. (1991). *The saturated self: Dilemmas of identity in contemporary life.* New York: Basic Books.
Gondra, J. M. (2013). A psychology of liberation for Central America: The unfinished work of Ignacio Martín-Baró (1942–1989). *The Spanish Journal of Psychology, 16*(e31), 1–17. https://doi.org/10.1017/sjp.2013.53
Gone, J. P. (2008). 'So I can be like a Whiteman': The cultural psychology of space and place in American Indian mental health. *Culture & Psychology, 14*(3), 369–399.
Gone, J. P. (2013). Redressing first nations historical trauma: Theorizing mechanisms for indigenous culture as mental health treatment. *Transcultural Psychiatry, 50*(5), 683–706.
González Rey, F. L. (2004). La crítica en la psicología social latinoamericana y su impacto en los diferentes campos de la psicología [Critique in Latin American social psychology and its impact in diverse psychological camps]. *Interamerican Journal of Psychology, 38*(2), 351–360.
Goodman, L. A., Liang, B., Helms, J. E., Latta, R. E., Sparks, E., & Weintraub, S. R. (2004). Training counseling psychologists as social justice agents: Feminist and multicultural principles in action. *The Counseling Psychologist, 32*(6), 793–836.
Gordon, T. (1973). Notes on White and Black psychology. *Journal of Social Issues, 29*(1), 87–95.
Grabe, S., & Dutt, A. (2015). Counter narratives, the psychology of liberation, and the evolution of a women's social movement in Nicaragua. *Peace and Conflict: Journal of Peace Psychology, 21*(1), 89–105.

Gutiérrez, G. (1988). *A theology of liberation: History, politics, and salvation* (15th anniversary ed.) (Caridad Inda and John Eagleson, Trans) (Maryknoll: Orbis, 1988; 1st ed., Maryknoll: Orbis, 1973). Originally published as *Teología de la liberación: Perspectivas [Liberation theology: Perspectives]* (Lima: CEP, 1971).
Guzzo, R. S. (2015). Critical psychology and the American continent. In I. Parker (Ed.), *Handbook of critical psychology* (pp. 406–414). New York: Routledge.
Hassett, J. J., & Lacey, H. (Eds.). (1991). *Towards a society that serves its people: The intellectual contribution of El Salvador's murdered Jesuits*. Washington, DC: Georgetown University Press.
Henriques, P., Hollway, W., Unwin, C., Venn, C., & Walkerdine, V. (1984). *Changing the subject*. New York: Methuen.
Herman, E. (1995). *The romance of American psychology: Political culture in the age of experts*. Berkeley, CA: University of California Press.
Hernández-Wolfe, P. (2011). Decolonization and "mental" health: A Mestiza's journey in the borderlands. *Women & Therapy, 34*(3), 293–306.
Hernández-Wolfe, P. (2013). *A borderlands view on Latinos, Latin Americans, and decolonization: Rethinking mental health*. Lanham, MD: Rowman & Littlefield.
Hoffman, D., Carter, D., Lopez, C.V., Benzmiller, H., Guo, A., Latifi, Y., & Craig, D. (2015). *Independent review relating to American Psychological Association ethics guidelines, national security interrogations, and torture [The Hoffman Report]*. Chicago: Sidley Austin LLP.
Hook, D. (2005). A critical psychology of the postcolonial. *Theory & Psychology, 15*(4), 475–503.
Jemal, A. (2017). Critical consciousness: A critique and critical analysis of the literature. *The Urban Review, 49*(4), 602–626.
Jenkins, J.H. (1991). The state construction of affect: Political ethos and mental health among Salvadoran refugees. *Culture Medicine and Psychiatry, 15*, 139–165. https://doi.org/10.1007/BF00119042
Jones, L. (1998). Adolescent groups for encamped Bosnian refugees: Some problems and solutions. *Clinical Child Psychology and Psychiatry, 3*(4), 541–551. https://doi.org/10.1177/1359104598034006
Jiménez-Domínguez, B. (1988). Una ubicación crítica de la psicología ambiental [A critical location of environmental psychology]. *Revista de Psicología de El Salvador, 7*(27), 7–23.
Jiménez-Domínguez, B. (2009). Ignacio Martín-Baró's social psychology of liberation: Situated knowledge and critical commitment against objectivism. In M. Montero & C. Sonn (Eds.), *Psychology of liberation* (pp. 37–50). New York: Springer.
Lassalle-Klein, R. (2014). *Blood and ink: Ignacio Ellacuría, Jon Sobrino, and the Jesuit martyrs of the University of Central America*. Maryknoll, NY: Orbis Books.
Leogrande, W. M. (1990). From Reagan to bush: The transition in US policy towards Central America. *Journal of Latin American Studies, 22*(3), 595–621.
Liem, R. (1988). El costo psicológico del desempleo: Estudio comparativo de hallazgos y definiciones [The psychological cost of unemployment: Comparative study of findings and definitions]. *Revista de Psicología de El Salvador, 7*(29), 213–241.
Liem, R. (1995). Psychology under fire. *Peace and Conflict: Journal of Peace Psychology, 1*(4), 441–444.
Lindorfer, S. (2007). *Sharing the pain of the bitter hearts: Liberation psychology and gender-related violence in Eastern Africa*. Berlin: Lit Verlag.
Lira, E. (2008). Psicología, ética y seguridad nacional: El rol de los psicólogos [Psychology, ethics and national security: The role of psychologists]. *Psykhe, 17*(2), 5–16.
Lykes, M. B. (1994). Terror, silencing, and children: International multidisciplinary collaboration with Guatemalan Maya communities. *Social Science and Medicine, 38*(4), 543–552. https://doi.org/10.1016/02779536(94)90250-X
Lykes, M. B. (1997). Activist participatory research among the Maya of Guatemala: Constructing meanings from situated knowledge. *Journal of Social Issues, 53*(4), 725–746. https://doi.org/10.1111/j.1540-4560.1997.tb02458.x

Lykes, M. B. (2012). One legacy among many: The Ignacio Martín-Baró Fund for Mental Health and Human Rights at 21. *Peace and Conflict: Journey of Peace Psychology, 18*(1), 88–95.
Lykes, M. B., & Moane, G. (2009). Whither feminist liberation psychology? Critical explorations of feminist and liberation psychologies for a globalizing world. *Feminism & Psychology, 19*(3), 283–297. https://doi.org/10.1177/0959353509105620
Lykes, M. B., & Sibley, E. (2014). Liberation psychology and pragmatic solidarity: North–South collaborations through the Ignacio Martín-Baró fund. *Peace and Conflict: Journal of Peace Psychology, 20*(3), 209–226.
Makkawi, I. (2015). Community psychology enactments in Palestine: Roots and current manifestations. *Journal of Community Psychology, 43*(1), 63–75.
Makkawi, I. (2017). The rise and fall of academic community psychology in Palestine and the way forward. *South Africa Journal of Psychology, 47*(4), 482–492.
Marín, G. (1991). Obituary of Ignacio Martín-Baró. *American Psychologist, 46*, 532.
Martín-Baró, I. (1986). Hacia una psicología social de la liberación [Toward a social psychology of liberation]. *Boletín de Psicología. 5*(22), 219–231 [English translation by Adrienne Aron in *Writings for a liberation psychology.* (A. Aron & S. Corne (Eds., Trans.)]. Cambridge, MA: Harvard University Press, pp. 16–32.
Martín-Baró, I. (1994). *Writings for a liberation psychology* (A. Aron & S. Corne (Eds.), Trans.). Cambridge, MA: Harvard University Press.
McCoy, A. (2006). *A question of torture: CIA interrogation, from the cold war to the war on terror.* New York: Metropolitan Books/Henry Holt.
Melville, M., & Lykes, M. B. (1992). Guatemalan Indian children and the sociocultural effects of government-sponsored terrorism. *Social Science and Medicine, 34*(5), 533-548.
Moane, G. (2003). Bridging the personal and the political: Practices for a liberation psychology. *American Journal of Community Psychology, 31*(1/2), 91–101.
Montero, M. (1988). Factores psicosociales en la elección de pareja [Psychosocial factors in the choice of a couple]. *Revista de Psicología de El Salvador, 7*(30), 337–349.
Montero, M. (1989). La psicología social en América latina: Desarrollo y tendencias actuales. *Revista de Psicología Social, 4*(1), 47–54.
Montero, M. (2004). Relaciones entre psicología social comunitaria, psicología crítica y psicología de la liberación: Una respuesta Latinoamericana [Relations between community-social psychology, critical psychology, and social psychology of liberation: A Latin American answer]. *Psyche, 11*(2), 17–28.
Montero, M. (2017). Psychology of liberation revised (a critique of critique). In B. Gough (Ed.), *Palgrave handbook of critical psychology* (pp. 147-161). London: Palgrave Macmillan
Montiel, C., & Rodriguez, A. (2009). Liberation movements during democratic transition: Positioning with the changing state. In C. C. Sonn & M. Montero (Eds.), *Psychology of liberation* (pp. 155–172). New York: Springer.
Montero, M., & Sloan, T. S. (1988). Understanding behavior in conditions of economic and cultural dependency. *International Journal of Psychology, 23*(5), 597–617. https://doi.org/10.1080/00207598808247788
Mpande, E., Higson-Smith, C., Chimatira, R. J., Kadaira, A., Mashonganyika, J., Ncube, Q. M., et al. (2013). Community intervention during ongoing political violence: What is possible? What works? *Peace and Conflict: Journal of Peace Psychology, 19*(2), 196–208.
Pacheco, G., & Jiménez, B. (Eds.). *Ignacio Martín-Baró (1942–1989) Psicología de la liberación para América Latina [Liberation Psychology for Latin America].* Guadalajara: Instituto Tecnológico y de Estudios Superiores de Occidente, Universidad de Guadalajara.
Painter, D., & Terre Blanche, M. (2004). Critical psychology in South Africa: Looking back and looking ahead. *South Africa Journal of Psychology, 34*(4), 520–543.
Paloma, V., García-Ramírez, M., Camacho, C., & Olmedo, L. (2016). How are we studying migrant well-being? A critical view from a liberating community psychology approach. *Community Psychology in Global Perspective, 2*(1), 73–86.

Portillo, N. (2012). The life of Ignacio Martín-Baró: A narrative account of a personal biographical journey. *Peace and Conflict: Journal of Peace Psychology, 18*(1), 77–87.
Quesada, J. (1998). Suffering child: an embodiment of war and its aftermath in post-Sandinista Nicaragua. *Medical Anthropology Quarterly, 12*(1), 51–73.
Robledo Gómez, Á. M. (2008). La formación de psicólogas y psicólogos en Colombia [Formation of psychologists in Colombia]. *Universitas Psychologica, 7*(1), 9–18.
Rose, N. (1998). *Inventing our selves: Psychology, power, and personhood*. Cambridge, UK: Cambridge University Press.
Sánchez, M. B., & Trujano Ruíz, P. (2011). Percepción de inseguridad publica y justificación de la violencia de estado en un grupo de habitantes del estado de México [Perceptions of public insecurity and the justification of state violence in a group of residents in the state of Mexico]. *Acta Colombiana de Psicologia, 14*(2), 35–43.
Sánchez, M. E., & Almeida, F. E. (1992). Synergistic development and participatory action research in a Nahuat community. *The American Sociologist, 23*(4), 83–99.
Sharim, D., Kovalskys, J., Morales, G., & Cornejo, M. (2011). Psychosocial trauma and memory: Designing a biographical framework to investigate the impact of the National Commission on Political Imprisonment and Torture in Chile. *Revista de Estudios Sociales, 40*, 81–88.
Singh, A. A. (2016). Moving from affirmation to liberation in psychological practice with transgender and gender nonconforming clients. *American Psychologist, 71*(8), 755–762.
Singh, A. A., Hofsess, C. D., Boyer, E. M., Kwong, A., Lau, A. S., McLain, M., et al. (2010). Social justice and counseling psychology: Listening to the voices of doctoral trainees. *The Counseling Psychologist, 38*(6), 766–795.
Sloan, T. (1988). Una visión critica de la personalidad. *Revista de Psicología de El Salvador, 7*(29), 243–251.
Sonn, C. C., Arcidiacono, C., Dutta, U., Kiguwa, P., Kloos, B., & Maldonado Torres, N. (2017). Beyond disciplinary boundaries: Speaking back to critical knowledges, liberation, and community. *South Africa Journal of Psychology, 47*(4), 448–458.
Sousa, C., & Marshall, D. J. (2017). Political violence and mental health: Effects of neoliberalism and the role of international social work practice. *International Social Work, 60*(4), 787–799.
Speight, S. L., & Vera, E. M. (2004). A social justice agenda: Ready, or not? *The Counseling Psychologist, 32*(1), 109–118.
Teo, T. (2015). Are psychological "ethics codes" morally oblique? *Journal of Theoretical and Philosophical Psychology, 35*(2), 78–89.
Tilly, C., Castañeda, E., & Wood, L. J. (2019). *Social movements, 1768–2018*. New York: Routledge.
United Nations High Commission for Refugees (2020). *Global trends: Force displacement in 2019*. http://www.unhcr.org/refugee-statistics
Wallace, B. C. (2000). A call for change in multicultural training at graduate schools of education: Educating to end oppression and for social justice. *Teachers College Record, 102*(6), 1086–1111.
Walsh, R. T., & Gokani, R. (2014). The personal and political economy of psychologists' desires for social justice. *Journal of Theoretical and Philosophical Psychology, 34*(1), 41–55.
Walton, J. R., Nuttall, R. L., & Nuttall, E. V. (1997). The impact of war on the mental health of children: A Salvadoran study. *Child Abuse & Neglect, 21*(8), 737–749.
Watkins, M., & Shulman, H. (2008). *Towards psychologies of liberation*. New York: Palgrave Macmillan.
Watts, R. J. (2004). Integrating social justice and psychology. *The Counseling Psychologist, 32*(6), 855–865.
Watts, R. J., Griffith, D. M., & Abdul-Adil, J. (1999). Sociopolitical development as an antidote for oppression—Theory and action. *American Journal of Community Psychology, 27*(2), 255–271. https://doi.org/10.1023/A:1022839818873

Whitfield, T. (1994). *Paying the price: Ignacio Ellacuría and the murders Jesuits of El Salvador*. Philadelphia, PA: Temple University Press.

Yen, J. (2016). Psychology and health after apartheid: Or, why there is no health psychology in South Africa. *History of Psychology, 19*(2), 77–92.

Zuñiga, R. (1975). The experimenting society and radical social reform: The role of the social scientist in Chile's Unidad Popular Experience. *American Psychologist, 30*, 99–115.

Toward a Decolonial Approach to Psychosocial Accompaniment from the "Outside"

Mary Watkins

> *If you come here to help me, you are wasting your time.*
> *If you come because your liberation is bound up with mine, then let us work together.*
>
> —Aboriginal Activist Group, Queensland, Australia, 1970's

Decolonial philosopher Nelson Maldonado-Torres (2008) confronts us with the fact that "(c)olonialism creates a reality in which some subjects become privileged givers while others do not even have bread to eat or to give" (p. 151), an "us" and a "them." How are we—as community psychologists, carrying the privileges we do—to create decolonial ways of working across this colonial binary when our day-to-day lives are implicated in the social structures that reproduce the need and suffering we address professionally?

As community psychologists, to responsibly answer invitations to enter into marginalized communities that are not our own in order to engage in common work, we need to deepen a process of self-reflection, of psychic and relational decolonization. As we know, this effort of "relational accountability" (Wilson, 2009) occurs in a historical time that is shaped by ongoing colonialism and coloniality. To begin, it is best not to think that it is possible to fully liberate ourselves from the imposition of colonial relationships, but that we need to maintain a vigilance for coloniality, an openness to critique, and a commitment to a path of trying to ever more deeply understand the interpersonal and intercommunity effects of our own seeing, understanding, and living. Alongside such understanding, however, de-powerment, de-privileging, re-distributing resources (land and wealth), and other actions for social and environmental justice are necessary correlates of relational accountability.

Our discipline—psychology—has its own colonial history, marked by its practices of evaluation, testing, diagnosing, researching *about* people, making interventions, pursuing treatments, and offering prognostications. Its history is marred by its

M. Watkins (✉)
Pacifica Graduate Institute, Santa Barbara, CA, USA

G. Stevens, C. C. Sonn (eds.), *Decoloniality and Epistemic Justice in Contemporary Community Psychology*, Community Psychology,
https://doi.org/10.1007/978-3-030-72220-3_6

use of the construct of the "primitive," its racist arguments, and its collusion with colonial values and forces (Brickman, 2003; Keller, 2007; Khanna, 2003). A decolonial approach to mutual accompaniment interrupts research *on* and proposes instead inquiry *by* and *with* those in a particular situation. It interrupts practices of diagnosing and treating "from above" and seeks to create an ethics of being alongside of, of learning together, of acting in solidarity and co-shouldering necessary risks. How are we to put such an ethic into practice?

Mutual Accompaniment

> ...the choice is between accompanying or not accompanying the oppressed majorities.... This is not a question of whether to abandon psychology; it is a question of whether psychological knowledge will be placed in the service of constructing a society where the welfare of the few is not built on the wretchedness of the many, where the fulfillment of some does not require that others be deprived, where the interests of the minority do not demand the dehumanization of all.
>
> —Ignacio Martín-Baró (1994, p. 46)

Liberation psychologist Ignacio Martín-Baró was speaking to those of us who have a choice—by virtue of our social privilege—to work alongside the majorities or to remain apart from them, as we pursue both psychological work and our private lives. Once we make that choice—for accompaniment—another type of work needs to begin: ongoing work to decolonize our minds, relationships, and theories and to face into the need for land and wealth reparations. For those psychologically-minded practitioners with socioeconomic privilege—often but not always white—the move to more horizontal practices of accompaniment is both needed and vexed. The learning is lifelong; the learning curve is steep, and the need for humility is great. This essay is an attempt to help us—I include myself, a white, upper-middle-class, professional, United Statesian—to get our bearings in decolonizing praxes of community and ecologically-based work. How do we shift our orientation and of what does such a shift entail?

Building solidarity with others to create beloved community—community where resources are equitably shared, and the dignity of all members is respected—is a slow intergenerational process. Beyond simply "showing up," it requires fundamental shifts in one's life priorities, values, and the risks one undertakes. If a person begins accompaniment from within a binary construct of "helper" and "helped," a conversion will be necessary to experience the potential and desirability of mutuality in accompaniment. Building solidarity together requires ongoing learning and continuing apprenticeship to cultural and ecological workers who are "inside" accompanists, community members themselves.

By "outside" accompaniment, I am drawing attention to psychosocial accompaniment engaged in by a person or group that does not belong to the group that has requested accompaniment. Outside accompanists who have privilege—that is, socioeconomic, racial, ethnic, citizenship, ableness—are faced with the additional enduring task of reflexively and critically examining their own positionality and developing a sensitivity of openness and responsiveness to critical feedback they receive from members of the community they are working with that better orients them in their relationships and work (see Watkins, 2019).

Shifting Orientations

> …speaking nearby or together with certainly differs from speaking for and about.
>
> – Trinh T. Minh-Ha, 1989, p. 101

Those of us who have access to excess social and economic privilege have choices. We can segregate ourselves in a very small corner of a sadly divided world, working to increase our own personal happiness and security, relatively unconcerned with the effects of our living on others; or we can take decisive steps to live in a more inclusive and challenging world where relationships across differences attune us to the lives of others and to the effects of our privilege and the systems that generate them, perpetuating injustice. In doing so, we can engage the struggle to understand how we can most meaningfully build solidarity with others to co-create a more just, peaceful, free, and sustainable world, and set about the work of doing so.

Reverse Osmosis: Stepping Out, Turning Toward

> The head thinks from where the feet are planted.
> —Brazilian proverb

When individuals and communities experience an extremely difficult situation—acute or chronic—some people from outside may come to "lend a hand." Others live in a psychic remove that pretends the difficulty does not exist, pretends that they are not implicated, and/or that they couldn't do anything about it if they wanted to. Privilege, including economic and white privilege, can operate to foster a remove from both knowing about and responding to the critical social and ecological realities of our time and their impacts on communities.

The insults and harms a community is dealing with can be redoubled by the absence of others, and by their failures of acknowledgment, empathic concern, and action. Accompaniment can be a needed—even if insufficient—antidote to the injuries caused by others' passive bystanding or active denial of the human suffering in their midst. While accompaniment cannot wipe away the pain born of traumatic injuries—individual or collective—it can begin to set into motion needed processes of psychic and social restoration through witness, support, and work to change the structural conditions that gave rise to the trauma. For this to even be a possibility, potential accompanists with privilege need to challenge habitual and normalized tropisms to turn away from the difficult. Often accompaniers with more relative privilege experience themselves as crossing from the familiar and the comfortable into the unfamiliar and the unsettling. They need to challenge the normalization of their own privilege so that they can see links that may exist between their privilege and the misery or difficulty they are walking toward. This reverse osmosis, of intentionally moving across boundaries in the opposite direction of normative responses that bystand or move away from others' difficulties, is a clear mark of accompaniment from the "outside" and is needed so that one can begin to think "from where the feet are planted," as well as to enable the heart to feel where the feet are planted. I want to clearly differentiate movement across borders that is committed to developing solidarity over the long haul from a voyeuristic and opportunistic movement of privileged people that suits their own curiosity or augments their own self-concept.

Horizontality: Coming Down to Earth and Moving Alongside of Others

For those born and educated into privilege, psychosocial accompaniment requires a fundamental re-orientation. Turning towards and moving alongside are necessary movements to orient the practice of accompaniment. Horizontality instead of verticality are the needed coordinates. This reorientation involves developing countertropisms: to look at rather than to look away, to listen rather than to speak prematurely, to encourage and respect others' leadership rather than always assuming and asserting one's own, to hang in rather than to flit away upon signs of difficulty.

Jordan Flaherty (2016), in *No More Heroes: Grassroots Challenges to the Savior Mentality*, describes the savior mentality as wanting to help others, but not being open to guidance from these very same people. The "savior" believes that others are helped more by him- or herself staying in the lead. The solutions saviors turn to are within the dominant system, even when that system is the problem, because the current system is the one that supports the privilege they do not want to interrogate. To go beyond merely performing horizontality, however, requires efforts to actively de-privilege oneself.

When accompanists are not a member of the community in which the work is unfolding, they are not on their own ground. They join others on theirs—even if this is a temporary place such as a refugee camp. The command accompaniers may wordlessly exercise in their own offices, neighborhoods, and classrooms quickly evaporates. They need to attune themselves to those around them, follow their lead, and lean into discerning how they might be of support to desired social transformations.

Being Invited; Being Asked to Stay Away

A person or a group may not be interested in accompaniment by a member from another group. A community may request that outsiders do not live near or even visit members. Being sensitive to whether or not one is invited and welcome is critical to an ethical practice of mutual accompaniment. There are very legitimate reasons for people and communities to refuse accompaniment or to be quite particular about how and where it is to occur, if at all, and by whom. For instance, a group may be clear that outsiders will take too much time and energy to educate and work with. They may have legitimate security concerns about infiltration by outsiders who seek to undermine their organizations. They may correctly perceive that it is the culture of the "outsider" that is creating the difficulty and so the outsiders' energies would be better spent within their own community, to transform it. Instead of educating "outsiders" one by one, a community may request that one work within already established networks of solidarity (Land, 2015).

In other instances, a group or community may issue a clear call for accompaniment or if offered accompaniment accept it with few reservations. For instance, international accompanists are often asked for in situations where a community perceives it is under threat. The community may discern that this threat needs to be brought to international attention and that having peace accompanists present may deter violent state or paramilitary incursions on the community. Asylum seekers caught in detention prisons and immigrants undergoing the threat of deportation may welcome accompaniment because they do not have adequate access to the resources they need to advocate for themselves.

Becoming Clear About What an Invitation is for

If an invitation has been extended, it is important to carefully clarify what it is for. There can easily be a collision between what the accompanier thinks in his own mind he has come to do and what is actually desired.

In "The White Savior Industrial Complex," Nigerian-American writer Teju Cole (2012) describes the harm that white people can do when they use "helping" to feel good about themselves and neglect engaging needed complex understandings of the situations they are sentimentally addressing. Speaking of Nigeria, Cole critiques white people who focus on

> hungry mouths, child soldiers, or raped civilians, [when] there are more complex and widespread problems. There are serious problems of governance, of infrastructure, of democracy, and of law and order.... Such problems are both intricate and intensely local.

The problems should not be separated from the foreign policies of white U.S. "helpers," and, in particular, from the United States' hunger for Nigerian oil, despite the human cost to Nigerians, other-than-human animals, earth, air, and water. Coles calls for "constellational thinking," arguing that racial privilege allows white people to deny the constellation of systemic causes in which they are complicit. Instead of indulging a sentimental need to "make a difference," Cole argues for "due diligence": doing no harm, proceeding humbly and with respect for the agency of the people around them. He cautions that activists and humanitarian workers need to become aware of how their actions may play into the hands of people who have more cynical motives. In particular, attention needs to be given to the role humanitarian aid and aid workers can play in sustaining the status quo by relieving immediate pressure on the need for systemic change.

Members of Indigenous Action Media (n.d.), a group that offers communications and direct support for Indigenous communities, offer a powerful critique of "allyship" in their article "Accomplices Not Allies: Abolishing the Ally Industrial Complex." It can easily extend to accompanists. They describe allies as most often providing only temporary support, too often romanticizing those they wish to "help," creating dependency, and positioning themselves as a savior. As an organization, they prefer to welcome those who are willing to be "*accomplices*," willing to commit a crime, to place themselves at-risk; a person who "has our backs."

> Accomplices listen for the range of cultural practices and dynamics that exists within various Indigenous communities.... Accomplices are realized through mutual consent and build trust. They don't just have our backs, they are at our side, or in their own spaces confronting and unsettling colonialism. As accomplices, we are compelled to become accountable and responsible to each other, that is the nature of trust.

We can also think of accomplices as co-conspirators, remembering that *conspirare* means to breath the spirit together.

One thing is clear: being "helpful" is not the application of one's arsenal of expertise regardless of the particularities of a given situation. Mary Pipher (2002), a clinical psychologist living in a U.S. city with a large influx of refugees, slowly realized that usual psychotherapy sessions were not what was most helpful to her new refugee neighbors. Sure, many had suffered trauma and needed someone attentive to this. But attention to this did not require doling out a psychiatric diagnosis but being a reliable and caring human being in the face of the egregious wrongs committed by other humans against this particular person and his or her group. Moreover, many refugees, are in more immediate need of learning how to navigate in their new environment so they can gain a foothold of security and familiarity. Rather than apply a lens of trauma to every refugee, accompaniers need to listen closely to where each person is and what it is they may desire assistance with.

Learning New Skills and How to Work Across Levels of Organization

People's needs and desires can be surprising and often they do not neatly fit the skill sets we have developed. We will often need to learn new skills or seek resources we may at first know little about. A group of Colombian social, political, and clinical psychologists from Pontifical Javeriana University have learned this firsthand. For the past 17 years, they have been engaged in a project of sustained psychosocial accompaniment, leaving the university in order to offer their support to people forcibly displaced by paramilitaries from the countryside to the capital of Bogatá. Many of these displaced persons have experienced acute and chronic violence and often the loss of family members (Sacipa, Vidales, Galindo, & Tovar, 2007). For themselves the accompanists are seeking to construct a daily practice that is consistent with their understanding of social commitment (Sacipa-Rodriguez & Montero, 2014). The members of this group—Social Bonds and Peace Culture—committed themselves to resist the trivialization of death and the rampant depersonalization of others that characterized daily reality in their society, a society that has been torn for decades by armed conflict.

Stella Sacipa-Rodriguez, one of the co-founders, describes her team's perspective on psychosocial accompaniment:

> …we conceive psychosocial accompaniment as a way of offering displaced people support and providing spaces for expressing and recognizing the emotional impact these violent events have had on them. Psychosocial spaces are designed to listen compassionately to the victims of forced displacement, spaces aimed at ensuring that these people feel accompanied, in order to provide conditions conducive to their recovery…
>
> Psychosocial accompaniment is a process marked by respect, acknowledgment of the human dignity of the person who has suffered displacement, a process which seeks to establish bonds and bridges for the renewal of confidence in a work of successive, respectful rapprochement, aimed at opening up the psychosocial relationship, to reach the heart of others from within oneself, through mutual recognition in everyday dialogue, in active listening and in shared work and play (Sacipa-Rodriguez, 2014, p. 67).

Psychosocial accompaniment often includes participatory research and other conscientizing efforts to construct "liberating knowledge," knowledge that will assist in transforming status quo arrangements that undermine the integrity of body and mind, relations between self and other, and between one community and another (Sacipa-Rodríguez, Tovar-Guerra, Galindo-Villareal, & Vidales-Bohórquez, 2009, p. 222).

> [It is] a process of offering the displaced person a space to recognize their emotional experience along with the possibility to express their feelings afterward, reflecting on the facts implied by violent acts. We speak of psychosocial process that facilitates recuperation and repair of social and cultural damage. We believe that accompaniment should be directed toward the affirmation of displaced persons as subjects in their own stories and the reconstruction of the social fabric of the community (Sacipa-Rodríguez et al., 2009, p. 222).

Through the collection of oral histories, the co-creation of support groups, the recognition and valuing of community resources that contribute to empowerment and resilience, the participants can create a community that gradually connects through ties of trust. It was in another arena, however, that the participants most desired accompaniment. Many of the displaced families wanted it to be clear in public records and memory that their loved ones were falsely assumed to be guerillas. They also wanted to know where their loved ones' remains are so that proper burials could be conducted. Honoring these deep desires, the psychologists needed to become knowledgeable about and effective in interfacing with relevant judicial and public authorities and processes.

This group emphasizes that a fuller recovery from such psychosocial suffering requires societal circumstances that make meaningful work, peace, and a dignified life possible. For the psychosocial reconstruction of a community to be ultimately effective, it must be part of a total approach that includes changes in the social, economic, and political life of the country. For these reasons, at a systems-level, the psychologists have also been exploring their possible contributions as psychologists to creating cultures of peace in Colombia.

Resisting Uncritical Representations and Stereotypes of Ourselves and Others

Critical race and Indigenous studies scholar and Aluet, Eve Tuck (2009), cautions us that spaces where oppression has occurred can become "saturated with the fantasies of outsiders" (p. 412). In accompanying, we need to struggle to become aware of both negative and idealizing stereotypes we bring that lead us to misread situations and to distort the personhood/being of another or others. Once we have discerned these stereotypes, we need to dismantle them, understand their history and functions, and refrain from projecting them. In idealizing projections, I experience either my ideal projections on another or chose only those aspects of their self-presentation that fit with my need to idealize. I avoid or am blind to the complexity and multifaceted nature of the one(s) I am idealizing. The latter was brought to my attention by several educators who work inside of prisons and who noted the tendency of some volunteers to valorize all those incarcerated, failing to appreciate the unique complexity of each person they were working with. Negative stereotypes that we project can have disastrous consequences. This is true for both human and trans-species relationships. bell hooks gives an example of this in her discussion of dominant cultures' representations of those who are "poor" as lazy, shiftless, and dishonest. In speaking of representations of "the poor," hooks (2006) cautions us to resist normative representations of others that diminish them before we have even had the chance to meet them (p. 248).

Mary Annette Pember (2017), in "This November, Try Something Different: Decolonize Your Mind," describes how many white people who seek her out as an Obijwe don't want to hear about what she hopes to share with them: "the diversity of Native cultures, the complexity and history of federal policies affecting us, our sophisticated understanding of our relationship to the earth, or the fact that Native peoples embrace popular culture in addition to their own traditions." Instead, they speak of Ojibwe in the past tense and want to know how they can participate in ceremonies, dances, drumming, and spirituality. "They wanted me to give them Indian names, identify their power animals, and teach them how to be shamans.

Mostly they just wanted to play Indian." Her white interlocutors too often stray into defining Ojibwe by their "plight" as seen by many white people. They also unreflectively superimpose idealizing representations, seeing Indigenous as "supernatural noble, selfless defenders of the Earth." Pember laments: "Our cultures, traditions, and spirituality have been subsumed into the great buffet of American consumerism; we are food for hipster and New Age appropriation, one in the dizzying blur of passing social memes." Aware of missionaries' critique of Ojibwe as wasting their time by speaking with one another rather than being more productive in a Western sense, she ironically proposes that decolonizing our minds needs to begin with just such speaking with one another, getting to actually know one another. This would help us to become aware of our stereotypes, representations, and idealizations.

Looking in the Mirror is Opening our Ears… and Keeping them Open

> Self-reflection is not a turn inward but a turn toward otherness.
> —Kelly Oliver, 2001, p. 219

There is much about ourselves that we can never understand with only an inward turn. We discover ourselves in and through our relationships, and particularly in relationships where we are open to differences and to feedback. As we try to better understand what decolonial accompaniment might look like, we have also been describing its potential shadow: intrusion, entitlement, disempowerment, projection, stereotyping, diminishing. It is best to name these potential pathologies that carry over colonial relations. Hopefully, the "accompanier" realizes that she is not the only one doing the looking, the observing. Some of the descriptors applied to would-be accompanists traveling under a "colonial passport" are "fly-by-night humanitarian workers," "trauma tourists seeking disasters by deserted beaches," "trauma trophy academics," "industrial white saviors." Some try to rebrand their vacations as generosity (Flaherty, 2016). Indigenous Action Media holds up this mirror to would-be allies: "parachuters," "essentially missionaries with more funding," members of the "ally industrial network" or "establishment," "co-opters," "savior allies," "action junkie tourists." These labels bespeak the harm created by too-frequent failure by would-be accompanists to take the time to deeply enough understand the situation being suffered, to examine their own relationship and the relationship of their country or group to the problem, and to make an enduring commitment to a community under stress. Instead frameworks and interventions derived from one cultural location are thoughtlessly applied to another location one knows almost nothing about. They also speak to the need for soul-searching about who such work is for, and whether the potentially self-serving nature of it has been disguised or minimized.

Hopefully, as accompanists we wonder how we are seen, and are willing to discover things about ourselves we never imagined, or only feared. Of course, what privilege we enjoy is not invisible to those we work with, far from it. By leaving our comfort zone, we may find that what we have taken for granted about ourselves and the shape of our lives is thrown into question. We may feel shame, guilt, and/or embarrassment. We risk a rupture of our certainties.

At the fragile moments of meeting others and in our efforts to reflect on them, we need to be attentive to what we bring in advance of actually experiencing one another. For here we find ourselves, as much—if not more—than we find who we are hoping to know more deeply. In accompanying each other as accompaniers, there needs to be space made for claiming and unpacking these moments.

bell hooks (2006) underscores that "awareness is central to the process of love as the practice of freedom" (p. 248). This love bears no resemblance to the sentimental. It is a *critical* love that identifies blind spots that perpetuate domination. When those we partner with experience us as open to their critical loving eyes, being "called out" can be transformed into being "called in." Indeed, embodying the capacity to accept and learn about oneself and others through critical feedback is necessary to deserving any trust in relationships across the differences constructed by injustice and racism. Meeting the challenge of divesting excess privilege and working through the dominance that is ingredient to coloniality makes it possible to slowly enlarge the circle of the "beloved community."

Jordan Flaherty (2016) counsels,

> And, if people from the community you are seeking to support give you negative feedback, if you are a person in a position of privilege and feel "called out," don't act defensively. Don't be fragile. Listen, learn, acknowledge, and use the experience as an opportunity to change your approach. (p. 199)

Cross-racial, cross-class, cross-gender, and cross-ethnic relationships can help white activists see themselves from others' points of view. It is important, however, that they are aware of how to work with their own potential responses of defensiveness, denial, and what Robin DiAngelo (2011) describes as "white fragility." She says that many white people when confronted about their racism or are criticized by people of color are quick to take offense, have their feelings hurt, or respond with denial or anger, rather than take in the learning that is being offered about their attitude and/or behavior. The work of outside accompaniment needs to be understood as including the work of opening oneself to such critique and mitigating against habitual responses that close down the potential learning necessary to be in authentic solidarity.

Reflexivity: Problematizing One's Own Identity, Positionality, Privilege, and Actions

Outside accompanists need to reflect on their own complex positionality with respect to dimensions such as race, ethnicity, class, gender, sexual orientation, nationality, religion, ableness. They need to reflect on how their complex positionality may contribute to perpetuating structural injustice. This entails facing into seeing how our positionality creates challenges and contradictions in our practice of accompaniment. What is the history between the groups we belong to and those we may be accompanying? How does it affect attempts to relate to one another, to understand each other, and to effectively act with each other?

Research on peace accompanists has clarified that outside accompanists, those not from the community inviting accompaniment, may have a very different narrative of why they have been invited from that of the members of the community who offered the invitation. An accompanist may think that both insiders and outsiders value the development of relationships across cultural differences and think that friendships are being formed. Those inside may think more practically about the value of having outsiders present, for instance as a deterrent to violence. The outside accompanist may not be as clear as the insiders that it is the relative value given to, for instance, a white body or a U.S. citizen over those in the community who are at risk that makes accompaniment desirable. This reality places the outside accompanist in an ironic position, deploying their racial and citizenship privilege even though they may be committed otherwise to dismantling such privilege.

Reflections on and research about protective accompaniment by those involved in it demonstrate the development of nuanced understandings that are possible through a commitment to ongoing reflexivity. Preliminary—often idealistic—ways of understanding self and others and one's actions as an accompanist give way to more realistic assessments that are truer to our complex identities in relationship to one another.

The practice of international protective accompaniment began in the 1980's when members of *Witness for Peace* and *Peace Brigades International* offered their presence in Nicaragua to help protect communities in danger from the U.S. supported war against the Sandinistas. Thousands of church members from the U.S. joined delegations to travel to Nicaragua and learn first-hand about the Nicaraguans' struggles for peace and justice. When they returned home, they formed part of a powerful lobby against U.S. foreign policy in Central America. Peace Brigades International began to accompany activists and civil society organizations operating in Guatemala, often under death threat (Mahoney, 2013).

Since the early 1980's approximately 1000 volunteers from 12 countries have accompanied human rights activists, refugees being repatriated from Mexico, and victim witnesses. Their presence is aimed at deterring violence by being able to make it known to the international community in ways that could result in sanctions or condemnations. Accompanists help to both increase the political action space for local activists to work in, while decreasing the impunity of local actors who threaten the activists' work (Mahoney & Eguren, 1997). Accompanists witness and report, making violent actions visible to civil society and the international community. Violent actors—paramilitary or state-sponsored—are usually not looking for witness of their atrocities and international condemnation. They tend to back away when there is an international presence. Other functions of accompanists include encouragement, solidarity, and the legitimizing of human rights activities. The daily life of accompanists can include "escorting human rights defenders; living in communities of displaced people; meeting with the military or civilian authorities; observing a demonstration, checkpoint, or road-block; launching an international alert; writing a report; or simply cleaning a house" (Mahoney, 2013).

Accompanists are careful to establish relationships with their congressional representatives, whom they rely on to help them communicate with generals in charge of a specific geographical location. Once accompanists return home, they are vocal against policies that support the ongoing violence. Their firsthand knowledge of the grassroots situation in a given country has the potential to make them powerful spokespersons for developing solidarity movements.

Aware of their excess privilege due to "First World" citizenship, and often because they are white, accompanists use the power accrued from their social privileges to help protect others who are deemed more expendable, and in danger due to their commitment to peace and human rights (Mahrouse, 2014). Gada Mahrouse highlights that in doing so, accompanists enter into a contradiction that is best claimed and reflected upon: by using privilege, one is an embodiment of the systems one hopes to be dismantling. Rather than resting in an idealistic representation of one's presence as benevolent, Mahrouse argues that such accompanists need to understand the paradox and contradiction of their actions and, if they decide to proceed, to understand that the deployment of privilege is a pragmatic decision and a compromising action. Instead of representing oneself as in solidarity with those accompanied, she suggests seeing one's accompaniment as principled pragmatism, using a troubled and troubling role to increase the safety and visibility of a given human rights struggle.

Etienne Roy Grégoire and Karen Hamilton (2016) adopt Foucault's understanding of reflexivity as thought that questions the very ground upon which one thinks, that attempts to discern where one is located within webs of power. Following Foucault, they differentiate between arterial and capillary power, between overarching power structures and diffuse relations of power in on the ground daily relationships. Using Foucault's analysis of power, they argue that accompanists protecting human rights activists imperiled by state actors are better prepared to understand the large-scale power dynamics, the arterial level, that they are acting within: i.e., the Guatemalan government, the international community, and the Inter-American Court of Human Rights. Whereas, when accompanists are involved in more local conflicts, the capillary level of diffuse power relations are difficult to discern and understand. No longer conducted on the stage of state power, accompanists find it hard to read the conflicting and contradictory identities of the people they are living among with regard to past and ongoing violence, and thereby have a limited understanding of how their presence is actually affecting a given situation. Grégoire and Hamilton (2016) suggest that accompanists take a more agnostic approach to these local situations in which they accompany so that they do not distort and misunderstand them. It is too easy to collude with oppressive local practices, including corrupt leadership, when one does not adequately perceive and understand them.

Such efforts at reflection and reflexivity do not necessarily end in accompanists from the outside going home, though sometimes this may be the case. Rather they make it possible to trade an untroubled sense of their own identity and privilege for a more realistic and nuanced one. This new view better enables them to understand how those they accompany experience their presence and their role. They become clearer about whether or not their potential to reinforce hegemony is outweighed by the immediate protective effects of their presence and the projects they are engaged in.

Accompanists may bring a practice with a laudable value into a context where it has unintended effects. Psychologists Dixon, Durrheim, Kerr, and Thomae (2013) found that dialogue opportunities between members of oppressed communities and members of oppressing communities, while potentially encouraging prejudice reduction, can de-energize resistance and protest by diminishing "the extent to which social injustice is acknowledged, rejected, and challenged" by those affected by it (p. 1). These scholars warn against the possible deleterious effects of pursuing elements of reconciliation such as prejudice reduction without prior or at least parallel success at achieving increased justice in areas such as racism, classism, and sexism. Often the word "reconciliation" is a misnomer, because there has not been an established relationship that can be restored (Watkins & Shulman, 2008). One is actually building relationships for the first time. When efforts toward building relationships between members of an oppressor group and members of an oppressed group occur, the caveat of the potential negative effect of such meetings on resistance needs to be acknowledged and efforts taken to mitigate against this possible downside. Indeed, Clare Land (2015) says that Palestinians' calls for "non-normalization" reject the building of relationships that give the "deceptive appearance of tolerance, democracy and normal life" because they "create the impression that harmony exists or that it can be created without attention to the return of land and refugees" (p. 251). The kind of mutual accompaniment that is sought is common effort to resist the Occupation for the sake of both Israelis and Palestinians.

From "Helping" and Cultural Invasion Toward Mutuality and Cultural Synthesis

Unfortunately, efforts at accompaniment can easily go awry if the colonial framework of "helping," "charity," and "being of service" are not thematized and deconstructed. Too often, humanitarian, community, and psychosocial work occur within the same structure of colonial relations that gives rise to a community's suffering in the first place. Hierarchical relations are mindlessly reproduced, ignoring or denigrating the knowledge of already marginalized community members. Interventions from the outside can displace and disable indigenous approaches better suited to the particular local context. Ameliorative actions can neglect the deeper causes of distress, particularly those of systemic injustice. When this occurs, creative and transformative work that could have emerged from processes of dialogue and collaboration across differences in experience and knowledge is thwarted. Disempowerment for community members prevails, while “solutions” fail in places that were not understood in advance of the application of knowledge derived from elsewhere.

Paulo Freire (2000) called this cultural invasion and carefully contrasted it with cultural synthesis. In cultural synthesis people “do not come to *teach* or *transmit* or *give* anything, but rather to learn, with the people, about the people’s world” (p. 180)—there is a synthesis of worlds through efforts of mutual understanding. In culturally invasive approaches, interpretations and interventions are imposed from the outside. Psychosocial accompaniment counters the cultural invasion of exporting to places around the world diagnoses, treatment interventions, and research agendas that should not be universalized and imposed from positions of falsely presumed cultural supremacy. The orientation of psychosocial accompaniment requires that we be mindful of the power of each individual and group to construct meanings and to transform aspects of their world. The creativity and resilience of those accompanied and their own cultural resources for understanding, healing (if needed), resistance, and struggling for structural change should be honored and supported, not usurped, in accompaniment.

Seeking Integrity of our Understanding with our Action: The Role of Reintegrative Shame

> Confronting, instead of quickly covering, an experience of shame as revelation of oneself and of society—facing “actual life”—requires an ability to risk, if necessary to endure, disappointment, frustration, and ridicule. Commitment to any position or to any loyalty, like commitment to another person, involves the risk of being wrong and the risk of being ridiculous. It is *relatively* easy to take even difficult action if one is sure one is right, that one has grasped the truth of a situation; it is relatively easy to entertain multiple possibilities of truth and of right action if one remains a spectator on the sidelines. Far more difficult than either is to give everything one is in supporting all the truth one can see at any given time, with full awareness that there are other possibilities and that further knowledge may enlarge and revise the hypotheses on which one has risked everything. Engagement with life and with history—self-discovery and further discovery of the world—has always involved just such risks.
>
> —Helen Merrell Lynd, 1961, p. 232

Accompaniment by “outsiders” is a risky endeavor and is best undertaken with an acceptance of likely failures and feelings of deserved shame. Shame, however, can be generative. I distinguish deserved from undeserved shame (Casey & Watkins, 2015; Watkins, 2018). Sometimes we are made to feel undeserved shame by things that happen to us, for instance, being raped, growing up in poverty, being disparaged for our skin color. At other times, we do or do not do something that causes us to feel grave disappointment in ourselves. Perhaps we did something with one intention, and find that grievous unintended consequences resulted and we are suffused with shame for our inadequate forethought or understanding.

Psychosocial accompaniment almost inevitably awakens feelings of deserved shame in “outside” accompanists, feelings that when metabolized have the potential for generating critical insight. This inevitability issues from what "psychosocial" means. The one who accompanies from the outside in a psychosocial manner has learned to see—to better understand—structurally the social misery they are witnessing. For instance, it becomes clear to them, as Jon Sobrino puts it, that "poverty results from the actions of other human beings" (Griffin & Block, 2013, p. 14).

The capacity to recognize feelings of deserved social shame is not, of course, an endpoint, but a gateway to reparative actions. These not only align one's evolving social understandings with one's daily living, enhancing the integrity of the accompanist, but are hopefully catalytic in reaching our mutual goals.

Decolonizing Lenses for Psychosocial Accompaniment

Decolonial lenses for psychosocial accompaniment must include a critical historical lens, a transformative lens (over and above an ameliorative approach), and a strength-based lens (vs. a damage and deficit-centered lens).

Historical Lens

Accompanists learn to see the situation they are addressing in its historical context. This is necessary so that distress observed in the present can be traced to its structural and historical roots. For instance, by seeing the growth in incarceration of people of color in the U.S. through the lens of history, Michelle Alexander (2010), author of *The New Jim Crow: Mass Incarceration in the Age of Colorblindness*, was able to understand it as one more expression of a racial caste system that has been present since the inception of the U.S. and its early embrace of slavery. The growth of the prison industrial complex during the "War on Drugs" was not about containing increasing rates of crime but reflected the use of the criminal justice system for racial control.

As we endeavor to create and embody decolonial psychosocial and ecopsychological practices, we must ground ourselves in an understanding of the history of colonialism and its present-day embodiment in neoliberal power and practices. We need to develop an understanding of psychology's complicity with colonial relations and practices. For instance, U.S. students need to understand the 500 years of worldwide colonialism that has brought us to this historical moment, while learning in detail about settler colonialism and slavery in the U.S., the effects of U.S. imperialism, and the development of internal colonies throughout the U.S. This historical understanding clarifies the relative failure of psychology to deviate from individualism as its rooting paradigm, and thus its failure to conceive of the individual within a fully delineated historical, psychosocial, and environmental context. A decolonial approach to community psychology must be a historical approach that includes and centers the experiences of the dispossessed.

Ameliorative and Transformative Lenses

Without understanding the wider historical context of the difficulties we witness, much of the work of helping professions consigns itself to ameliorative approaches. Critical community psychologists Nelson and Prilleltensky (2005) have clarified differences between ameliorative and transformative approaches. While both are concerned with promoting well-being, they often do so differently. The ameliorative approach, as they define it, focuses on various interventions at the personal and relational levels, emphasizing values of holism, care, and compassion. The transformative approach engages people in a collective level of analysis, seeking to understand and effect the root causes of affliction. It focuses on eliminating oppression and unjust power differentials, while embodying and enacting values of social justice, interdependence, collaboration, egalitarianism, and solidarity. The transformative approach strives for collective and systemic change through collaborative partnerships and community participation. In their intervention processes these partnerships aim for "conscientization, power-sharing, mutual learning, resistance, participation, supportive and egalitarian relationships, and resource mobilization" (Nelson & Prilleltensky, 2005, p. 145).

When enhanced well-being is understood through the lens of power, possible outcomes delineated by Nelson and Prilleltensky (2005) are the following:

> …increased control, choice, self-esteem, competence, independence, political awareness, political rights and a positive identity; enhanced socially supportive relationships and participation in social, community, and political life; the acquisition of valued resources, such as employment, income, education and housing; and freedom from abuse, violence and exploitation. Outcomes at multiple levels of analysis that emphasize power-sharing and equity are in the foreground. (p. 159)

Psychosocial accompaniment includes both attention at the individual and relational levels, but, like transformative approaches, seeks to understand and address systemic causes of suffering. Through seeing with a wider historical lens, psychosocial accompanists do not reduce the difficulties a person faces to the individual level of analysis, but neither do they neglect the needs of individuals and communities who are presently afflicted. Refugee communities, for instance, may seek accompaniment to get help meeting immediate needs. At the same time, such accompanists may be involved in advocating at policy and legislative levels for refugee rights while supporting national policy interventions that could reduce the need for forced migration.

Strength-Based Versus Damage and Deficit-Centered Lenses

Unfortunately, human service agencies and providers, as well as community psychologists, have too frequently peddled damage and deficit-centered narratives of the people and communities with whom they work. In addition to the racism that can be an ingredient to such narratives, there also may be self-serving and self-perpetuating motives, conscious and unconscious. As they are currently configured, "helping" professions are dependent on addressing the "needs" of their clients. Seeing others principally through their needs is a highly distorting lens. Such a lens constructs the other in a disempowering light, even if the stated goal is empowerment. "Helping" unwittingly becomes complicit with harming.

In reflecting psychologically on community and ecopsychological accompaniment, we need to become clearer about the lens(es) we are using to represent others and ourselves and their effects. For instance, Eve Tuck (2009) urges us to become aware of the long-term effects of damage-centered research on marginalized populations. While such studies, she says, have been used in the hope of obtaining needed reparations and political or material gain, it is time to consider the long-term consequences of communities thinking of themselves as broken and being presented to others uni-dimensionally in this light. Damage-centered studies reduce those studied to speaking of their pain, deprivation, and wounds. When the colonial context of these wounds is not clearly articulated, those who have suffered the wounds are themselves seen through a pathologizing lens, neglecting a focus on the cultural and individual pathologies of the perpetrators. As feminist liberation psychologist Geraldine Moane (2000) underscores, the strengths born of oppressive conditions, such as generosity, courage, perseverance, ingenuity, and solidarity are too infrequently adequately acknowledged and witnessed.

Tuck (2009) contrasts the pathologizing consequences of a damage-centered paradigm to the empowering and affirmative aspects of a paradigm focused through the lens of desire. While not denying the loss and despair wrought by colonization and ongoing racism, a focus on a community's desire points members toward the future they hope is possible and the efforts underway to achieve that future. It creates a space for both vision and hard-won wisdom. Tuck concedes that there is a role for damage-centered inquiry to document the harms suffered, but, she says, communities are so much more than the damage they have had to endure. Accompanists need to struggle to avoid both denigrating and idealizing attitudes and assessments, as they try to understand a particular community in a balanced and realistic light.

Mutual Accompaniment as a Pathway to Solidarity and Co-Liberation

> Solidarity is an uneasy, reserved, and unsettled matter that neither reconciles present grievances nor forecloses future conflict.
>
> —Eve Tuck & K. Wayne Yang (2012, p. 3)

> I don't believe in charity; I believe in solidarity. Charity is vertical, so it's humiliating. It goes from top to bottom. Solidarity is horizontal. It respects the other and learns from the other. I have a lot to learn from other people. Each day I'm learning. *Soy un curioso.* I'm a curious man.
>
> —Eduardo Galeano, *Barsamian* (2004), p. 146)

Some practices of accompaniment can be criticized as still laced with coloniality, exhibiting a condescending paternalism that fails to understand the mutual and reciprocal possibilities in accompaniment. Accompaniment may be, however, a path on the way to more decolonial relationships and work in solidarity. If we patiently penetrate the taken-for-granted "normality" of oppressive social structures and their realities, one discovers that those on the "outside" suffer in their own way—not to the same extent—from what those on the "inside" of an oppressed community are suffering from. We discover that our liberations are interdependent. At that point, accompaniment is experienced as mutual accompaniment. Often those who begin by thinking of themselves as involved in "service" discover that it has worked in "reverse"—that they have been transformed, have learned, and may have realized their own spiritual poverty. Such experience directly undermines one-sided narratives of helping. Mutual accompaniment is bi- or multi-directional and of benefit to each partner— not in the same ways, not from the start, not immediately—but through commitment to the long haul of individual, social, and community transformation.

With an intentional and practiced generosity, feminist Chicana writer Gloria Anzaldúa (1987) worked against a politics of exclusion and toward a visionary inclusivity. For instance, she invited white people in, knowing that in time—as she put it—they would find that that they are not "helping," but "following our lead" (p. 107). The accompanier becomes accompanied. Anzaldúa argued that Anglos and Chicanos are "implicated in each other's lives," that they leak into each other, taking on the attributes of one another. "So we are really neither one nor the other; we are really both" (p. 243). She practiced "calling in" when it is possible. Radical relationship coach Mel Mariposa (2016) distinguishes calling in from calling out, describing calling out as a one-sided declaration and critique of another regarding how they have fallen short that serves little hope for their change. It has a performative aspect that can be shaming and outcasting. In contrast, calling in both holds one accountable and invites one to do better. It rejects a reactive dismissal of someone with some shared values and encourages active listening to build stronger mutual understanding.

Solidarity does not depend on a homogeneity of identity or shared origins. It depends on shared purpose, desire, vision, and action. It entails sharing equitably the work and risk of actions, often disproportionately borne by those with less privilege and power. Collaborative solidarity, as discussed by trans- and Black Lives Matter activists Elle Hearns, Aaryn Lang, J Mase, and Kei Williams, assists the most marginalized to lead the work. It asks people to reflect on how they show up and centers communities to speak for themselves. This is both a demand and an ethics directed at those of us who are involved in accompaniment from "outside." Tuck and Yang (2012) underscore that solidarity is not an endpoint; it is never finally achieved. It is not a resting place but a process where conflicts and challenges rightly emerge. These conflicts and challenges need to be anticipated and taken on as an essential part of the work of psychosocial accompaniment.

References

Alexander, M. (2010). *The new Jim Crow: Mass incarceration in the age of colorblindness*. New York: The New Press.

Anzaldúa, G. (1987). *Borderlands/La frontera*. San Francisco: Aunt Lute.

Barsamian, D. (2004). *Louder than bombs: Interviews from the progressive magazine*. Boston: South End Press.

Brickman, C. (2003). *Aboriginal populations in the mind: Race and primitivity in psychoanalysis*. New York: Columbia University Press.

Casey, E. S., & Watkins, M. (2015). *Up against the wall: Re-imagining the U.S.-Mexico border*. Austin, TX: University of Texas Press.

Cole, T. (2012). The white-savior industrial complex. *The Atlantic*. Retrieved from https://www.theatlantic.com/international/archive/2012/03/the-white-savior-industrial-complex/254843/

DiAngelo, R. (2011). White fragility. *International Journal of Critical Pedagogy, 3*(3), 54–70.

Dixon, J., Durrheim, K., Kerr, P., & Thomae, M. (2013). 'What's so funny 'bout peace, love and understanding?': Further reflections on the limits of prejudice reduction as a model of social change. *Journal of Social and Political Psychology, 1*, 239–252. Retrieved from https://jspp.psychopen.eu/article/view/234

Flaherty, J. (2016). *No more heroes: Grassroots challenges to the savior mentality*. Chico, CA: AK Press.

Freire, P. (2000). *Pedagogy of the oppressed*. New York: Continuum.

Grégoire, E. R., & Hamilton, K. (2016). International accompaniment, reflexivity and the intelligibility of power in post-conflict Guatemala. *Journal of Genocide Research, 18*(2–3), 189–205.

Griffin, M., & Block, J. W. (Eds.). (2013). *In the company of the poor: Conversations with Dr. Paul Farmer and Fr. Gustavo Gutiérrez*. Maryknoll, NY: Orbis Books.

hooks, b. (2006). *Outlaw culture: Resisting representations*. New York: Routledge.

Indigenous Action Media. (n.d.). *Accomplices not Allies: Abolishing the Ally Industrial Complex*. Retrieved from http://www.indigenousaction.org/accomplices-not-allies-abolishing-the-ally-industrial-complex/

Keller, R. C. (2007). *Colonial madness: Psychiatry in French North Africa*. Chicago: University of Chicago Press.

Khanna, R. (2003). *Dark continents: Psychoanalysis and colonialism*. Durham, NC: Duke University Press.

Land, C. (2015). *Decolonizing solidarity: Dilemmas and directions for supporters of indigenous struggles*. London: Zed Books.

Lynd, H. M. (1961). *On shame and the search for identity*. New York: Science Editions.

Mahoney, L. (2013). *The accompaniment model in practice*. Retrieved from http://archives.forusa.org/fellowship/2013/summer/accompaniment-model-practice/12385

Mahoney, L., & Eguren, L. E. (1997). *Unarmed bodyguards: International accompaniment for the protection of human rights*. West Hartford, CT: Kumarian Press.

Mahrouse, G. (2014). *Conflicted commitments: Race, privilege, and power in solidarity activism*. Montreal, QC: Mc Gill-Queens University Press.

Maldonado-Torres, N. (2008). *Against war: Views from the underside of modernity*. Durham, NC: Duke University Press.

Mariposa, M. (2016). A practical guide to calling in. *Consent 101*. Retrieved from https://theconsentcrew.org/2016/05/29/calling-in/

Martín-Baró, I. (1994). *Writings for a liberation psychology*. Harvard University Press.

Minh-Ha, T. T. (1989). *Woman, native, other: Writing post-colonialism and feminism*. Bloomington, IN: Indiana University Press.

Moane, G. (2000). Making connections: Systems of resistance. *Irish Journal of Feminist Studies, 6*, 221–226.

Nelson, G., & Prilleltensky, I. (2005). *Community psychology: In pursuit of liberation and well-being*. New York: Palgrave Macmillan.

Oliver, K. (2001). *Witnessing: Beyond recognition*. Minneapolis, MN: University of Minnesota Press.

Pember, M. A. (2017). This November, try something different: Decolonize your mind, *Yes!* Magazine. Retrieved from http://www.yesmagazine.org/people-power/this-november-try-something-new-decolonize-your-mind-20171113

Pipher, M. (2002). *The middle of everywhere: Helping refugees enter the American community*. New York: Houghton Mifflin.

Sacipa, S., Vidales, R., Galindo, L., & Tovar, C. (2007). Psychosocial accompaniment to liberate the suffering associated with the experience of forced displacement. *Universitas Psychológica, 6*, 589–600.

Sacipa-Rodriguez, S. (2014). To feel and to re-signify forced displacement in Columbia. In S. Sacipa-Rodriguez & M. Montero (Eds.), *Psychosocial approaches to peacebuilding in Colombia*. New York: Springer.

Sacipa-Rodriguez, S., & Montero, M. (Eds.). (2014). *Psychosocial approaches to peacebuilding in Colombia*. New York: Springer.

Sacipa-Rodríguez, S., Tovar-Guerra, C., Galindo-Villareal, L. F., & Vidales-Bohórquez, R. (2009). Psychological accompaniment: Construction of cultures of peace among a community affected by war. In M. Montero & C. Sonn (Eds.), *Psychology of liberation: Theory and applications* (pp. 221–236). New York: Springer.

Tuck, E. (2009). Suspending damage: A letter to communities. *Harvard Educational Review, 79*(3), 409–427. Retrieved from http://pages.ucsd.edu/~rfrank/class_web/ES-114A/Week4/TuckHEdR79-3.pdf

Tuck, E., & Yang, W. (2012). Decolonization is not a metaphor, decolonization: Indigeneity. *Education and Society, 1*(1), 1–40.

Watkins, M. (2018). The social and political life of shame: The U.S. 2016 presidential election. *Psychoanalytic Perspectives, 4*(2), 25–37.

Watkins, M. (2019). Mutual accompaniment and the creation of the commons. Yale University Press.

Watkins, M., & Shulman, L. (2008). *Toward psychologies of liberation*. New York: Palgrave Macmillan.

Wilson, S. (2009). *Research is ceremony: Indigenous research methods*. New Brunswick, Canada: Fernwood Publishing.

Decoloniality and Participatory Action Research in Puerto Rico

S. Miranda Gierbolini

"There is no other, we must embrace Hope"

Nelson Torres Santos

Puerto Rico is a colony of the United States. On June 16th, 2015, the United Nations' Special Committee on the Situation of Implementation of the Declaration on the Granting of Colonial Countries and Peoples, urged the USA to create the necessary conditions for the full application of Resolution 1514 (XV) of the United Nations Assembly (2015). This committee has repeatedly approved resolutions on behalf of the people of Puerto Rico. Before the USA invasion in 1898, the Spaniards occupied the island practically annihilating the native Tainos by means of genocide and ethnocide. Since the Spanish conquest in 1593, Puerto Rico has been a settler Colony.

Trías Monge (1999) has referred to Puerto Rico as the oldest colony in the world. This in reference to the Spanish Empire invasion in 1493 and the 1898 occupation of the United States of America until this date. In 1952, the Congress of the United States of America approved our constitution, which grants Puerto Rico the status of Free-Associated State referred to as Commonwealth of Puerto Rico. As a result, the United Nations withdrew Puerto Rico from the list of colonies. As of August of 1967, the Decolonization Committee approved Puerto Rico's inclusion among the colonies of the world. The Commonwealth of Puerto Rico is considered a farce, since USA Congress did not grant full self-government to the people of Puerto Rico. Bolton (1953) stated that the Federal Government will do for Puerto Rico as it does for the other 48 states of the Union. It is not to intervene in any issue which normally is not reserved for a federal government in a Federal system. Furthermore, affirmed that it is an agreement which cannot be broken or abrogated by any one of its parts.

S. M. Gierbolini (✉)
University of Puerto Rico, San Juan, Puerto Rico
e-mail: Dolores.miranda@upr.edu

G. Stevens, C. C. Sonn (eds.), *Decoloniality and Epistemic Justice in Contemporary Community Psychology*, Community Psychology,
https://doi.org/10.1007/978-3-030-72220-3_7

In summary, Puerto Rico holds limited self-government powers irrespective of having its own Constitution approved by the Congress of the United States. This Constitution is subdued to the Constitution of the United States. All local judicial-political and economic decisions are subdued to the powers of USA Congress. Furthermore, the 1917 USA Cabotage or Coastal Laws remained the same having an aversive impact on their economy. Puerto Ricans do not vote for the Congress nor the President of the United States, even though their decisions affect them.

The past decade has been a time and space in which the public debt and financial crisis in Puerto Rico has exacerbated the tensions between the Puerto Rico and U.S. The situation is critical and has been accompanied by alternatives such as not paying the debt, question it in court locally or negotiate which have brought out in the open the colonial status of the Puerto Ricans. The Congress of the USA has assigned a Financial Supervisory Board to oversight all dealings with regard to the debt. This Board overpowers local government imposing conditions that attempts against the life and violate human rights of the Puerto Ricans. This decision has made clear the double bind position of local government in which the colonial status is part of the problem and an impediment to its solution. While the people of Puerto Rico have made efforts to assert economic and political practices, the restraints of the colonial and territorial status has become more intense and thus has created the urgency to address a solution to the US-Puerto Rico colonial relationship.

The issue of coloniality is in continuous debate, there appears to be a general understanding that the existing political status has created an assistance or welfare psychology which has impeded the people to affirm self-management efforts. Taken this understanding as a consensus, it is possible to develop a parallel socio economical experience at the community level aiming to create a counter discourse to dependency and the reproduction of the colonial being (Maldonado, 2007). This requires a collective engagement in which various levels are *addressed: the coloniality of power (domination and power), the coloniality of knowledge production and the subjectivity or lived experience of colonization* (Maldonado, 2007). A major point of consideration of this perspective is the fact that the coloniality of being is constructed by coloniality of power and coloniality of knowledge. Even when a decolonization process has taken effect, the *coloniality of being* surfaces. Fanon (1986) has referred to Black Skin-White Masks relation as the metaphor of the master-slave relation—an illustration of embodied coloniality.

Coloniality is constructed by modern globe order, a way of being which is a rationalized power that continues to produce violence (Adams, Dobles, Gómez, Kurtiş, & Molina, 2015). The human action of genocide, ethnocide and democide (Najenson, 1981) has characterized the colonization events throughout the globe. It has taken place in different historical moments and continues to prevail in diverse formats. Decolonization as a self-affirmation of the people of different nations under the dominance of the economic and political more powerful, is among us occupying the psychological terrain challenging traditional and mainstream epistemology, ontologies and praxis.

Consider that as recent as the late twentieth century many territories and nations of Africa, Asia, Caribbean, United States, and Europe were decolonized. Yet, some of the liberated countries encountered domination of the settlers as in the case of Puerto Rico (previous Spanish occupation), Canada, New Zealand, Australia, and South Africa, among others. Puerto Rico has had separatist movements both armed and political. People of Puerto Rico resisted the occupation of both the Spanish in 1593 and the United States in 1898. These efforts have taken different political forms as liberation struggle from political domain of United States and capitalism. The over 500 years of colonialism has been reinforced through all cultural ideological apparatus (Althusser, 2003), as well as of brutal and subtle repression. The modern episteme is embodied and imbedded in the deepest corners of our souls. Yet, no matter how modernized or how we reproduce the "American way" it will never be enough to create as equals, the same.

This chapter will address an academic endeavor from a critical psychology and decoloniality perspective. The interpretation of decolonization has had its shifts from the classic geopolitical perspective. The ongoing subjectivities as part and residual of colonization has a long-term effect. Considering this perspective, we will approach the colonial condition of Puerto Rico within a decolonial perspective as proposed by authors such as Mignolo (2008), Grosfoguel (2007), and Maldonado (2007). Fanon (1986) elaborated on the subjectivity of the colonized as he described the psychological appropriation of and reproduction of the oppressor's way of being, a process which transcends persons, time and space. This endeavor is an ongoing process which has crossed disciplines, psychologies and geographical borders. As such, it has been a journey through various theoretical shifts and paths.

The Beginning

The 1980s was a turning point in grassroots community efforts in Puerto Rico. We have placed our attention on grassroots communities referring to "the majority of ordinary people who form the mass of citizens of a hierarchically structured society, lacking any exceptional social advantages or social power" (Campbell, 2011, pp. 12–14). The liberation political efforts of the 1960 and 1970's left deception and a sense of learned helplessness, about decolonization. Nationalism which was a strong discourse during the national liberation efforts was reprocessed as part of an academic debate. It continues to be invoked by the different political status proponents (pro-statehood, commonwealth and separatist) participants, particularly when a crisis strikes. Democracy defined through electoral practices has provided a sense of empowerment, which strengthens the dependency on political parties. People were able to negotiate their needs with politicians promises. The people's hope and demands were music to the ears of politicians.

The 1970s generated an irreversible capitalism crisis due to the control over oil, the technology, state and capitalism triad (Hardt & Negri, 2000) which initiated different stages of mass unemployment and the pauperization of government. Industrialization faced major changes that reshaped the workspace in what was to be referred to as post-Taylorism or the postindustrial era (Hardt & Negri, 2000). The welfare-state initiated changes with regards to its responsibilities and services. By the 1980s it became noticeable that the welfare state was going through a major transformation. The new face of capitalism gained strength in the neoliberal regime in which labor benefits in the social and workplace scenarios were beginning to fade as the result of a precarity discourse. A new labor subjectivity was emerging in a neoliberal state context. By the mid-1980s, grassroots communities had organized and began questioning Government's privatization turn, specifically it is reducing responsibilities in the areas of basic services such as education, health, and employment.

At the public university, the Social Community Psychology program at the University of Puerto Rico was initiated in 1975. The program had a firm commitment to social change and justice. As an academic program, it addressed the epistemological and ontological crisis debate in psychology (López, 1988). There was a paradigm shift towards social constructionism and praxis outside the typical academic and agency scenarios. Community outreach and community organizations were the new scenarios. Knowledge production was not exclusive to the academic community. In addition, communities were not a setting to merely apply academic knowledge or science but, a space of joint knowledge production. The challenge was and is how to address poverty, social inequality and disparities in the very spaces these conditions take place. This meant to leave the comfort zone of the university, engaging in activities that generate research as part of its processes. With a group of students, I created the Social Community Workshop as a space to direct efforts to grassroots community organizations.

The grassroots community organizations that approached me, had taken a critical position concerning academics and the theories they use. From that position, the challenge was not of being, rather of becoming. The shift to critical community psychology (Coimbra et al., 2009; Teo, 2015) placed the political upfront on our agenda. The language of empowerment, embraced by Community Psychology, lost its meaning and, was contested due to the reflection with regard to power. Power was precisely not the sole goal of social change. A participative alternative was to become and did not seem congruent with the power language. The question was, how not to reproduce the power structure which was part of the problem? This concern was raised by the women who questioned power as an issue relevant to male patriarchy, not to the struggle for social justice and equality (Miranda, 1995).

The work of Freire (2005), Fals Borda (2001), Martín-Baró (1989), and critical theory have been important theoretical background. The social movement literature and more important the praxis in Central and South America, South Africa, Palestine, and Native Americans, among others, have contributed to the route we followed. It was clear, from our stance, that doing Social Community Psychology with grassroots communities was not possible without a decolonization and decoloniality agenda. The challenge was to legitimize it within the academic and professional psychology hegemony. We have continued our journey; receiving different reactions, yet gaining strength and recognition among colleagues. Outstanding leaders such as Antonia Pantojas and Francisco Figueroa affirmed that the purpose of the community endeavor was to liberate our country and ourselves. It would be possible if people were engaged in a critical analysis of the socio-historical construction of their subjectivity, and participate in an alternative social and economic experience and collective activism.

Several authors have conceptualized social movements with interest in those referred to as alternative or popular (Camacho and Menjivar, 1989; Castells, 1981; Fals Borda, 1985). This literature recognized new forms of social movements and how to distance themselves from the practices of class movements that dominated until the seventies. Camacho and Menjivar (1989) states that activation in civil society seeks changes in political society, but not necessarily to assume power. Popular movements express the interests of those who, contrary to the hegemonic groups, are not intended to maintain the dominant social structure. He notes that reforms may be local, regional, class-based, pluralistic or specific. They can operate in a disarticulated and discontinuous mode, however, they can become structured movements and political purposes. Reichman and Fernandez (1994) offer the following characterizations summarizing the literature on this topic: a) have an emancipatory orientation; b) are a compromise between those whose orientation is toward power and those whose orientation is towards cultural; c) they have a relatively anti-modernist perspective and do not share the linear view of history; d) social composition is heterogeneous; e) objectives and action strategies have differentiated thematic interests which are combined into a sociopolitical force preserving the autonomy of the different movements; f) organizational structure is decentralized and nonhierarchical; g) a politicization of everyday and personal private life, and h) generate unconventional methods of collective action. Much of the community organizations' activism in Puerto Rico have incorporated some conceptual elements of social movements due to their political activism.

The different conceptual orientations reviewed seem to agree that community movements: (1) Are initiated as an organized collective action towards a common goal following the structural contradictions of societies, (2) have an identification process where opposing social sectors address their differences, and (3) develop a social transformation alternative project that transcends its immediate objective. From this reference, we could interpret that community movements in Puerto Rico are organizing civil society about production issues (economic projects, everyday life (e.g., housing, health, education etc.) and symbolic or subjective such as cultural resistance. The state is identified as the adversary and experience has led them to propose alternative social organizations and social policies. It is important to recognize their political direction and position to exercise critical citizenship (Pizarro, 2013). We agree that this requires informed collective citizen participation aimed at creating the conditions of a society capable of assuming independence from the powers of the state, political parties and capital. Education plays a fundamental role in its critical and liberating forms as a cultural device. We assume a critical vision of citizenship that recognizes the contradictions and the violence in the exercise of their rights. As a participant noted: "Communities must have their own forms of direct democracy totally autonomous from the political parties. Dependency on government must be substituted with proposal and actions that permit the development of community politics".

The grassroots community organizations, which we most related to, are skeptical about social movements due to the history a centralization and assuming leadership over the local efforts. The local grassroots organization at the municipal and county localities claim for autonomy due to their conceptualization processes and positions. Therefore, the tendency has been to encourage voluntary alliances and join forces in actions issue related and for organizational purposes.

Decoloniality

Decoloniality or decolonization as a research theme or methodology in psychology does not have a prolific history judging by the scarce publications in psychology journals. Decoloniality and decolonization in psychology as a research topic and praxis may be of local interest to nations with a history of colonization subjugation. Apparently, it has not made it to the approval of the dominant psychologies that oversight what is published in their journals. This disinterest comes as a surprise considering the fact that over half of the countries in the world live the aftermath of colonization or continue to be a colony as is the case of Puerto Rico.

Recently, psychology scholars have taken an interest in decolonization because of its mental health and ethical implications (Blanco, Blanco, & Díaz, 2016). Others have addressed decolonization within the liberation and decolonizing psychology (Adams, et al., 2015). Blanco et al. (2016) have taken a structural approach alerting psychologists working with trauma, of the psychological underpinnings which become invisible while working from an individual perspective. They adopt the concept of disorganized order (Martín-Baró, 1989), referring to how society has developed ideologies with their corresponding actions, in the construction of social classes. Social classes are not a static data or a fixed reality, but they are groups, whose configurations depend, every moment, on the relations between groups involved in the economic, political process (Blanco et al., 2016).

The psychosocial trauma has been addressed by Sotero (2006) and Kirmayer, Gone, and Moses (2014). This perspective has been applied to analyze the long-lasting effects of the subjugation of one population to a dominant group as has been the case of colonization. According to Sotero (2006), subjugation requires physical and psychological violence, segregation or displacement, economic and cultural dispossession. This subjugation process has implied a psychological trauma for the affected populations. These forms of subjugations have persisted in the form of structural violence. Another conceptual approach is that coined by Alexander (2004) referred to as cultural trauma. Cultural trauma refers to when members of a collective have been subjected to an event that leaves indelible marks in the group conscience. Cultural trauma has common explanations with historical trauma. Cultural trauma does not require a historical event and might be a current experience (Stamm, Stamm, Hudnall, & Higson-Smith, 2014).

The trauma literature has the risk of getting trapped in a pathology perspective, with all the stigma and deviation attached to it. Yet, it does alert us to the adverse psychological consequences of colonization. More so, to the long term and reproduction of trauma for future generations. Once it forms part of the psychology agenda there is a chance to understand better the phenomenon, how to endure, as well as liberate from it. On the other hand, the trauma manifestations are a way of life of the coloniality being, without the pathological and risk implications. Although, there are judicial and legal implications in terms of human rights and claims for indemnity, as the case of political prisoners, the disappearance of the insurgents and resistance.

Decoloniality as an Intersectional Affair

Given that the colonial being is a social construction and not a natural event, then we can change the course of history. This entails engagement beyond our discipline and geographical boundaries. A culturally anchored critical psychology approach (Reyes Cruz & Sonn, 2011) is needed to uncover the structural, historical, economic and political underpinnings of subjectivity. A problem-orientated transdisciplinary approach is required while inquiring with uncertainty on firm ground. It requires placing upfront values, wisdom, and worldviews to create a common language, promote critical consciousness, de-naturalization knowledge conducive to collective action (Phillips, Adams, & Salter, 2015). Students, professors, and community participants contribute to the understanding of the problems addressed, developing new skills and knowledge. Scientific, artistic and popular knowledge from different disciplinarian formation creates a common ground. This basic collective and diverse base of knowledge sets the ground to develop socio-economical-political alternatives within a self-management community perspective.

Communities have engaged in a series of community efforts over the last 40 years. These efforts have evolved into what has been referred to as a community movement. Academics from a critical perspective, have espoused on grassroots community movements as ground for social change or transformation (López & Serrano Garcia, 1979; Miranda, 1995; Seda, 2012). This perspective has developed from the socio-historical realities that our country has experienced and is essential to articulate alternative proposals to the marginalization and dependency our colonialism status has subjected us to in our own territory. Grassroots community movements have positioned themselves in various ways with regard to these proposals at local and international levels. The social transformation and power relations ruptures within and outside community organizations situate us in *the political*. Alvarado, Botero, and Ospina (2010) conceptualize politics as the organization and governance of human communities. According to these authors, it involves various forms of government and social organizations associated with power relations between human beings. The *political* constitutes subjectivity, as it is lived or embodied, with its institutional and structural aspects (Alvarado et al., 2010). Our analysis of decolonization is based on these intertwined preliminary conceptualizations.

The Social Community Psychology Program scholars and students are approached by various community grassroots organizations. We function as a work-group, in which academic and community leaders constitute a joint effort to contribute to a collective political subjectivity, which can be part of a decolonizing process. We see decolonization as a process of deconstructing the subjectivities mediated by a history of colonialism, capitalism and political liberalism. In the work-group, we concur that academics and community leaders can contribute to and be part of a collective political subjectivity that results in a decolonizing process. Our focuses are on the critical methodologies of work and research that have emerged and are relevant to our purposes. Ruptures with the status quo methodologies are a necessary means to give the participants, in our case the community, a voice to tell their story, rescue their history, relate it to and share with other decolonized contexts, and work out strategies to change history. Tuhiwai-Smith (2012) suggests that research methodology, in a way, determines your questions, the tools and methods used, as well as shapes the analysis. Tuhiwai-Smith (2012) points out that for this reason, it should reflect the language, forms and interests relevant to indigenous epistemology. The research process should be informed by understanding the ways indigenous cultural groups perceive reality, their ways of knowing, and their value system (Chilisa, 2011). We emphasize that these should be the underlying premises of the method to use to recognize and understand the social subjectivities of the grassroots community actors if we want to agree on a decolonizing agenda.

The question that concerns us is how to contribute to and understand the process of building a political-ethical alternative that emerges from a collective political subjectivity. The process of decolonization construction engages alternative subjectivations that make a difference, challenging history. For a psychosocial approach, we find the conceptualization of Butler and Athanasiou (2013) on the dispossession useful. According to Butler and Athanasiou (2013), dispossession has two instances: one in which the subject comes into existence when installing itself lost with social norms that regulate their willingness to address other objects. The other instance is to be deprived:

> refers to processes and ideologies by which people are (un) owned and hopeless by regulatory and normalizing powers defining cultural intelligibility and regulating the distribution of vulnerability: loss of land and community, appropriation of one's body by others, as in the case of slavery, military restraint, imperial, economic violence, poverty, security systems, bio-political subjectivation, liberal possessive individualism, neoliberal governments and casualization" (p. 2).

Butler and Athanasiou (2013) emphasize that the process of change involves rupture and suffering for those who have been displaced and colonized, a painful condition imposed by the regulated violence that determine the terms of subjectivity, survival, and life-a condition which could be part of the picture of abandonment, the shattered lives and strenuous struggles, speeches that evade ethical and social responsibilities of fulminant damages, which are sometimes imperceptible (Povinelli, 2011).

Our work-group has addressed dispossession and abandonment proposing its reprocessing by affirming their sense of others and placing their pain and struggles in plain view, to deconstruct and de-normalize their self. The hope is that community activation could articulate a political subjectivation that could generate an alternative social project for the country. The proposition is a continuous doing and undoing process, which to a certain point, enriches experience recognizing its complexity. Social activism is constituted by multitudes (Hardt & Negri, 2000) and identities that have the challenge of recognizing themselves in order to generate alternatives from the plurality and identities in which they coincide. In the past, authors Miranda, De Leon, Franco, and Quiñónez (1991) and Serrano-García and López-Sánchez (1991) have agreed that this mobilization had not yet, been able to articulate a collective political proposal that fosters such social change but was in process.

Scholars have recognized the role of the academy in the constitution of critical citizenships (Pizarro, 2013; Miranda, Pizarro, & Santos Torres, 2013; Flores, 2011; Teo, 2015) contributing to alternative social projects and decolonization. Formal education is seen as a governance device (Foucault, 1979) that contributes to building political subjectivities. As such, education is interpellated to un-naturalize the colonial being. As scholars, we have had before us the challenge of assuming the responsibility that this role implies.

At present, the discourses of economic precariousness prevail and, as a result, rentable disciplines are privileged, sacrificing or letting die those that are not. The answer has been privileging profitable disciplines as Povinelli (2011) states, letting those that are not, sacrificed. It is precisely those disciplines that assume a critical stance that are at risk as they are recognized as obstacles to projects directed to create university profitable enterprises. The institutional policies assume an ethic that takes distance from collaboration and justice to be competitive in a world vision of the survival of the fittest. In this debate, critics reaffirmed as part of citizenship. This citizenship challenges the academics to construct theory from action, closing the distance between theory and practice, if there ever was one.

Decoloniality, as described earlier, is a conceptualization that suggests action to address the colonial geopolitical context, the knowledge, and the subjectivity triad. All three concepts constitute decoloniality. We have tried to understand the situation in Puerto Rico contributing to create alternative subjectivations—the process of creating alternative subjectivities in the social construction of a decolonial being. A liberated being. If this is possible, we must act upon the political colonial situation, make shifts and ruptures from the modernist academic theory and practice to construct alternative world visions and practices among us. We suggest creating collective scenarios for alternative education, a participative approximation in political organization outside political parties, economic solidarity practices and dispossession of the coloniality being.

Case Study

Within this overall picture, our work-group has collaborated with grassroots community movements in poor communities from a community self-management perspective. Community self-management refers to the constitution and functioning of communities based on autonomy, about the people's decision-making. Community self-management is elaborated within the Participatory Action Research paradigm. In this sense, people have the autonomy to initiate alternatives for their community. It is taking distance from political parties as to the conductor of their destiny. It is not to be interpreted as an individual action since it exists in relation to others and Solidarity. Martinez Diaz (2013) has expressed that community self-management in our context is referred to as a channel through which human infinite potentials are directed towards their goal with the solidary support of others. It proposes an alternative to the fatalistic vision of poverty, as the sum of deficiencies, towards a vision of hope. It is a process capable of liberating communities from dependency and margination to create a space of self/presentation, defense, and coordination with others. Community self/management is the *hope* of conquering economic, social and political for collective benefit or common welfare. It is a long-term project, therefore sustainability is indispensable (Martinez Diaz, 2013). It is a long-term project; therefore, sustainability is indispensable. It is fundamental values are democracy, liberty and solidarity (Martinez Diaz, 2013).

About the Participatory Action Research Method

Participatory Action Research has been the foundation of our work-group. Other approximations such as indigenous methodology (Chilisa, 2011) and decolonizing methodology (Tuhiwai Smith, 2012) have been incorporated. Participatory Action Research (PAR) proposes several possibilities among which we highlight two: a method for the study of community self-management and thus contribute to the accumulation of knowledge, and a method that contributes to performativity (action) policymaking (RedCimas, 2015). Systematization (Holliday, 2012) is an instrument of community self-management that produces elements relevant for reflections as well as research. Systematization is considered part of the community action-research. It refers to the critical interpretation of one or various experiences. They serve as the starting point of their reordering and reconstruction, discover or explains the logic of the process lived in them: the diverse factors that intervene, how they relate among themselves and why was it done in this way (Holliday, 2012). It is best referred to as Experiences Systematization. This Experience Systematization produces knowledge and significant knowledge, which enables the appropriation of the meaning of the experiences, theoretically understand them and orient them to a future with a transformative perspective. Reflection is an important assumption when addressing social change and community action (Flores, 2011). While the academic participants become part of the participatory process, their knowledge is shared in a relational knowledge production (Gergen, 2009) endeavor with the community. It is this relational effort that contributes to close the gap between popular and academic knowledge, producing new knowledge that will guide all participants (Fals Borda, 1985). Research is embedded in the actions which may have initiated.

Our accumulated experience suggests that a diversity of methodological approaches, make up a social tapestry of the research we proposed. It suggests a Do It Yourself approach (Kinchloe, 2005) which extends the spectrum of research and action in a dialogic continuous development. The continuous making and unmaking enrich the experience of recognizing its complexities. As Retamozo (2006) suggests, the research approaches to social activism, in our case through community activism, challenges us to diversify our research practices. There is no method or a single methodology that fits our purpose. The research approach is not linear and a sequential set of tasks. It is rather a method as position and in horizontal motion (Retamozo, 2006) interweaving stories, narratives, photos, profiles, audiovisuals, documents, files, conversations, situational analysis, forums, speeches and reveal political subjectivations.

Participants

The present case study is of a non-profit and non/governmental organization known as the Socio-economic Development Initiative of the Jobos Bay (SDI). This organization brings together grassroots community organizations of poor residents in rural and urban areas of Puerto Rico. Currently, its members are an urban organization in the metropolitan area, one urban organization from a southeast zone, and four from a mixed rural/urban southeast area. Each is dedicated to the social and economic development of their respective communities. While each has their history, they have formed what we refer to as a collective (colectivo) for mutual support under an umbrella organization. Communities from the metropolitan area are poor with a high incidence of police intervention and harassment. Their efforts are directed towards youth, providing after school activities, which include academic tutoring, sports and arts with local resources. They have a community theatre company, a community tee shirt business and a leadership development education program. The other organizations address environmental justice, collaborative and traditional fishing, alternative organic agriculture, solar energy, community gardening, afterschool education for children, community organization, social problems, ecotourism, and mangrove and manatee protection.

Our entry to the organizations was at the requests of the communities and as collaborators. We had connected in various activities within the environmental and community movements realm. The leaders approached us as collaborators and currently our Social Community Workshop (SCW) is considered part of their organization. Previous we worked with the initial organizations in an alternative education project which had as its objective to provide opportunities for high school degree completions and occupational training. This school was a formative scenario to provide work experience through community enterprise which evolved within the school. We had kept in touch participating in educational and social activities as well as providing support and companionship to their citizen activism.

Ten years ago, we were invited to collaborate in a new endeavor, which comprised the integration of various community organizations in the center and southeast of Puerto Rico. Together with my graduate students, I engaged in educational activities with community organizations in formation. We initiated workshops within participatory methodology (RedCIMAS, 2015) and popular education. The SCW has developed modules integrating the perspective of participative methodologies drawing from the work of Paulo Freire, Fals Borda, and Holliday.

It is typical for some communities to request a need and resource assessment. These are developed with the community leaders or interest groups in consultation with the community. Graduate students serve as facilitators and at specific moments students from the BA program participate. For example, students assisted in designing and collecting community data as requested by the community organization or community members. All data are analyzed and discussed with the community organization. "We participate in the design, questionnaire and data collection. Children and youth are part of the team and responsibilities are distributed among neighbors" (Adneris, community leader). After the data presentation, the community decides strategies and the activities that should follow. The SCW collaborates with these activities. Students decide their participation considering their formation and interest. SDI coordinator Roberto has expressed: "the community survey is a picture of the community. You look at the overall picture of where we are. It is an act of listening, more than talking. Listening is an important tool to know where we are, where to go and how."

Oral history of the community has been another introductory activity. This is an important activity considering it is an encounter between the older members with children and youth. The children and youth participate as interviewers and collect memorabilia such as tools, photos, songs, toys, games, recipes, etc. The presentation activates the community while rescuing their history, practices, cultures and world visions. It is a space and chance to have those who have migrated elsewhere, come and participate with their offspring.

Community organizations may also consult to assure participation of the community in a specific project, for example, to develop and strengthen ecological practices as recycling, sharing appliances, tools, etc. Another is a community garden and installation of solar energy in the community center as initial step of the solar energy community. We also facilitate networking with other academic programs. For example, if there is a need for a consultation in architecture, the corresponding academic program is contacted. There is also a community organization resource network. Other activities address the sense of community, planning social and educational activities at the community level. The Manglar Festival is an activity, that rotates around the different communities promoting discussions about social justice, recent community activism resisting and denouncing governmental and business alliances that threaten the environment and local community economic self-management, local cultural practices, strategies to protect the mangrove. These activities generate income for both the individual community organization and the Socio-economic Development Initiative of the Jobos Bay. They create a time and space to share traditions and ancestors' knowledge and practices, nature and sense of community practices. These festivals are a scenario in which the decolonizing methods projects are presented. For example, an oral history video of the community, dance, songs, intergenerational storytelling. Local artists participate with the youth giving workshops on various topics related to local trade.

The educational component provides group process facilitation, participative education training, community organization, and sustainable community business development. The systematization sessions provide a space for reflection, critical structural analysis of the work done, lessons and future actions. Once a month, time is taken to analyze the current, social, economic and political situation with regards to their organizations. These are referred to as conjunctural analysis (structural analysis). There are two monthly meetings: solidarity economy committees and the health and education committee. The other is the SDI assembly.

An important activity is the meeting with other community organizations in Puerto Rico to share a concern and agree on action. These meetings are a space to articulate a position in relation to various concerns and issues affecting the community. In assembly, positions are articulated within a participative methodology format (RedCimas, 2015). The decisions are elevated to the general population, politicians, local and international organizations. Participating organizations share their positions related to values such as life, justice, critical citizenship, solidarity economy and decolonization.

Recently, the Formation with Communities School was initiated. The TSC presented a curriculum and manual with the participation of community leaders. This school will provide workshops on diverse themes. It is based on popular education and participative methodology. It is another scenario where participants share their experience applying the knowledge imparted with the facilitator. The modules include the knowledge developed by participants. A weekly radio program has been initiated, funded by the communities. The radio program serves the purpose of giving voice to the communities contributing to their ongoing analyses, knowledge, and proposals addressing various themes and concerns. As Roberto has stated, "we want the people of the communities to share with others what has been done and learned".

Lessons

Community learning is the result of collective reflections of the experiences and actions from which new knowledge or reprocessed knowledge emerges. We addressed them as lessons. Trust and friendship are the foundation of the relationships between the academics and the community. This creates expectancies as such. Tradition values friendship as an asset and bond forever. Respect and loyalty prevail irrespective of social or personal differences. Years after his anthropological work with the cane workers, Mintz (1984) reflected on his relationship with his participants. After not seeing each other in 35 years Taso commented "there is the house, it is empty, I miss you". Mintz (1984) valued this friend to whom he gave a voice, yet he knew Taso had chosen him. When accepting to participate in his research there was a friendship bond to last a lifetime. In our case, they have commented, "you are not a collaborator, you are part of us". The role of academic research is secondary. Research provides a forum in which all voices be heard. The knowledge production occurs via many formats such as photos, narrative, videos, essays, thesis, dissertations, presentations, etc. This is a shared endeavor. When students and professors request permission for their research, it is considered a matter of reciprocity. Research is also part of the ongoing participatory action. The projects are not ours. There is no ownership pretense as is the usual consequence of the intervention of academic experts. Intellectual property is irrelevant and looked down upon because knowledge production is for the common good, not for personal lucrative.

Addressing poverty and environmental justice has provided routes to strengthen a sense of being collective. This is the foundational ground of our history. Issues of individualism rise at different points of the journey, particularly concerning responsibility, solidarity and economic solidarity emerges. The community enterprises are a collective endeavor, and some feel they contribute more and merit more retribution. On the other hand, some feel that once they have learned enough, and they should go on their own. Collectivity takes more time as decisions are collective and the benefits are for the whole. Some cannot afford to wait. These issues are addressed through critical reflection agreeing on practice adjustments.

Power or empowerment is not an objective. More so it is not relevant. This would mean reproducing the ways of the oppressor. Equity and respect are fundamental. It is not a matter of persuading or having influence, more so, the aim is to do together with our differences. Yet, this is not to be interpreted as raw relativism. The power and quest for power are irrelevant when striving for a liberated self. Nachito reveals that "I believe that if I had political power I could do things for my community. On the contrary, I was trapped in politics which responded to political parties' agenda and not my community".Communities assume a community self-management politics with regards to government, political parties and capitalism. Propositions emerge as new ways to resist capitalism and modernity ways: engaging in solidarity economy, reciprocity in efforts, fair distribution and care for nature.

Solidarity is an important foundation. It refers to being able to be with others beyond differences. It is the dynamic of caring for others in their process of growth and prosperity Eating in the community is part of our cultural tradition. Sharing food is considered an act of being with others. In Puerto Rico sharing your food is an act of caring and retribution. It is a moment of celebration in community and friendship. There is also a practical reason related to this act. When your garden or are dedicated to agriculture, the produce grown must be shared so it is spoiled. Especially, when there are not enough means of storage. In celebration, we share recipes, new forms of cooking and harvest.

Political participation varies among participants. There is participation in the typical electoral process. Yet, as time goes by, this participation is questioned and decreases. For some, their hopes are still expected to be reached through typical politics as a negotiation arena. An alternative political route is proposed. Through participative methodology, positions as a group are presented at the typical political, public and private spheres. The alternative route (of anarchist characteristics community activism) is to generate agreements to be implemented by the communities solving for themselves and establishing alliances with other communities, universities, NGOs, and businesses. In alliances, there are a time and service sharing agreement. This alternative is a big challenge due to dependency on government and political party history.

Oral history is an activity in which different generations learn the ways of ancestors. It is a means to also honor the elder and recognize their wisdom. The stories are presented in different formats for all ages and preferences. The narratives and get together contribute to rescue our history. That history modernity despises and made us ashamed of our origins. Restore self and community value. In addition to oral tradition, ironically the world vision of the past are the ways to save the planet Yet, there have been transformations and generations that will continue to in habitat between borders of the past, modernity and new present. The challenge is to move on to a new history.

Back to the Beginning

Our critical analyses suggest that we have experienced transformation in our practice. In the first phase from a social movement perspective, two positions were sustained: (1) demands toward the state for services in the area of health, education and environmental protection, (2) decolonization from a pluralist perspective. The second phase efforts were directed to develop solidarity economy. A third phase was directed to strengthen sustainable social economic projects. Currently, there are appearances of collective subjectivations and deep ecologism. The opportunity to experience a community self-management endeavor builds a sense of collective being which serves as a foundation for a liberation endeavor at the personal, local and national spheres. Community learning, with its lessons, contributes to enhancing social action towards a common goal, and the inter-relationality of humans, society and nature. The near 25 years of a university-community relationship has resulted in lessons, challenges and strategies that suggest a shift to decoloniality conceptual framework is pertinent. Our results are supportive of the critical tridimensional perspective of coloniality. There is much to investigate, understand, systemize and theorize. As an ongoing process decoloniality could form part of political action to which Social-Community Psychology can contribute from a transdisciplinary and international effort. We are convinced that knowledge produced in collaboration transforms social practices, politics, and everyday life.

Experience is a means to create a new discourse that counteracts and resists the prevailing coloniality being. Iñiquez-Rueda (2003) has advocated for counter-discourse as a strategy for social change. Community Psychology has a chance to have a protagonist role in the processes of decolonization and the social-historical justice implicated. The shift to critical community psychology and liberation psychology (*psicologia de liberacion*) has gained its place among academics and practitioners. As well as *psicologia de liberacion* with more than half the world living in the aftermath of colonization. Colonization still prevails in the world in different formats and expressions. Countries such as the United States has its classic colonies such as Puerto Rico and Native Americans in its territory. The story is repeated across the globe. Given this reality, there is what we call an urgent song for decoloniality practices. They are difficult tunes to follow as individuals, but as a group, it's a hit.

The traditional hegemonic theories and practice of Community Psychology are limited in making a difference. Our invitation to the critical is to give a place for the voices of the people with whom we work to compose the tapestry of an ongoing knowledge in our quest of being human. This endeavor crosses academic disciplines and geographical borders. Those of us who have lived the colonization know what it means to move between borders of oppressed and oppressor (Anzaldua, 2009; Fanon, 1986; Mignolo, 2008). This is not unique to our experience. As historical subjects, we are decentered moving in different directions transiting diverse subjectivations. Black Face, White Masks as per Fanon's metaphor is illustrative of the many paths of being Black and White. These can be a hybrid, new or uncertain. Decoloniality is like a journey to the Nepantla where:

> Bridges span liminal (threshold) spaces between worlds, spaces I call Nepantla, a nahuatl word meaning tierra entre medio. Transformations occur in this in-between space, an unstable, unpredictable precarious, always-in-transition space lacking clear boundaries. Nepantla es tierra desconocida and living in this luminal zone means a constant state of displacement, an uncomfortable, even alarming feeling. Most of us dwell in nepantla so much of the time it's become a sort of "home". Though this state links us to other ideas, people and worlds, we feel threatened by these new connections and change they engender (Anzaldua, 2009, p. 243).

Where are We Going?

We will continue to transit in the Nepantla, our journey, with new ideas we do not completely understand, but with others, they will make sense. Freire (2005, p. 47) wrote "The new man, then, will not be an oppressor, as assumed wrongly by the oppressed, instead of man liberating himself". In other words, following Freire, the new subjects are the oppressed struggling for their liberation. If so, it is the oppressed who have the responsibility to initiate the decolonizing process of emancipation. Gaztambide Fernandez (2012) has elaborated on solidarity as a key aspect of the oppressors liberating relationship with the oppressed. Thus, solidarity entails a collective project that requires dialogic participation, critical consciousness of the power relationship between the oppressor and oppressed. Solidarity in relation to decolonization is about what it means to be human (Gaztambide Fernandez, 2012). We are made in and through others, solidarity works to reveal new horizons, new ways of being together (Gaztambide Fernandez, 2012). Decolonization has multi-routes and unknown possibilities. Our struggles continue in the hope of a world yet to be. It is a tough journey, it hurts and is worth the pain….

References

Adams, G., Dobles, I., G6mez, L. H., Kurtiş, T., & Molina, L. E. (2015). Decolonizing Psychological Science: Introduction to the Special Thematic Section. *Journal of Social and Political Psychology, 3*(1), 213–238. https://doi.org/10.5964/jspp.v3i1.564.

Alexander, J. (2004). Toward a theory of cultural trauma. In J. C. Alexander, R. Eyerman, N. Giesen, S. Smelser, & P. Sztompka (Eds.), *Cultural trauma and collective identity*. Berkeley, CA: University of California Press.

Althusser, L. (2003). *Ideologia y aparatos ideológicos del estado*. Buenos Aires, Argentina: Nueva Vision.

Anzaldua, G. E. (2009). In A. L. Keating (Ed.), *The Gloria Anzaldua Reader*. Durham, NC: Duke University Press.

Alvarado, S. V., Botero, P., & Ospina, H. F. (2010). *Subjetividades políticas: sus emergencias, tramas y opacidades en el marco de la acción política. Mapeo de 61 experiencias con vinculación de jóvenes en Colombia* [Political Subjectivities: Their Emergence, Plots and Opacities in the Framework of Political Action. Mapping of 61 Experiences Linked to Youth in Colombia]. *Utopía y Praxis Latinoamericana, 15*(50), 39–55.

Blanco, A., Blanco, R., & Díaz, D. (2016). Social (dis)order and psychosocial trauma: Look earlier, look outside, and look beyond the persons. *American Psychologist, 71*(3), 187–198.

Bolton, F. (1953). *The case of Puerto Rico in the United Nations. Declaraci6n de la Honorable Señora Frances P. Bolton, Representante de los Estados Unidos ante la Cuarta Comisión sobre Puerto Rico*, el *30* de octubre de 1953.

Butler, J., & Athananasiou A. (2013). *Dispossession: the performative in the political*. Polity.

Camacho, D., & Menjivar, R. (1989) *Los movimientos populares en America Latina* [Popular movements in Latin America]. Siglo Veintiuno.

Castells, M. (1981) *Crisis urbana y cambio social* [Urban crisis and social change]. Siglo XXI.

Campbell, C. (2011). The role of collective action in the prevention of HIV/aids in South Africa. In K. Ratele, N. Duncan, D. Hook, N. Mkhize, P. Kiguwa, & A. Collins (Eds.), *Self, community and psychology*. Cape Town, South Africa: UCT Press.

Chilisa, B. (2011). *Indigenous research methodologies*. Los Angeles: SAGE Publications.

Coimbra, J., Duckett, P., Fryer, D., Makkawi, I., Menezes, I., Mountain, I., et al. (2009). Community critical psychologies: Ideological convergence despite geographical dispersion. *Journal of Critical Psychology, Counselling and Psychotherapy, 9*. ISSN 1471-7646.

Fals Borda. (2001). Participatory (action) research in social theory: Origins and challenges. In P. Reason & H. Bradbury (Eds.), *Handbook of action research* (pp. 27–37). London: Sage Publications.

Fals Borda, O. (1985). *Conocimiento y poder popular* [Knowledge and popular knowledge]. Siglo XXI.

Fanon, F. (1986). *Black skin, white masks*. London: PLUTO Press.

Flores, J. M. (2011). *Psicologia y praxis: una vision latinoamericana*. Cuernavaca, Mexico: Editorial Latinoamericana.

Foucault, M. (1979). Discipline and punish: The birth of the prison. (Trans A. Sheridan). Vintage.

Freire, P. (2005). *Pedagogia del oprimido* (2nd ed.). Mexico: Siglo XXI Editores S.A.

Gaztambide Fernandez, R. (2012). Decolonization and the pedagogy of solidarity. *Decolonization: Indigeneity, Education and Society, 1*(1), 41–67.

Gergen, K. (2009). *Relational being: Beyond self and community*. Oxford: Oxford University Press.

Grosfoguel, R. (2007). El giro decolonial: reflexiones para una diversidad epistémica más allá Del capitalismo global. Castro Gomez, S. & Grosfoguel, R. (Compiladores) Bogotá: Siglo del Hombre Editores.

Hardt, M., & Negri, A. (2000). *Empire*. Cambridge, MA: Harvard University Press.

Holliday, O. J. (2012). La sistematización de la experiencia, práctica y teoría para otros mundos posibles [Systematization of experience, practice and theory for possible worlds]. Centro de Estudios y Publicaciones Alforja

Iñiquez-Rueda, L. (2003). La psicología social como crítica: continuismos, estabilidad y efervescencias: tres décadas después de la crisis. *Revista Interamericana de Psicología, 32*(2), 221–238.

Kinchloe, J. L. (2005). On to the level: Continuing the conceptualization of the bricolage. *Qualitative Inquiry, 11*(3), 323–350. https://doi.org/10.1177/1077800405275056

Kirmayer, L. L., Gone, J. P., & Moses, J. (2014). Rethinking historical trauma. *Trancultural Psychiatry, 51*(3), 299–319.

López, M. & Serrano García I. (1979). Hacia una psicología social-comunitaria [Towards a Social-Community Psychology]. Unedited document of Área de Psicología Comunitaria:Universidad de Puerto Rico, Río Piedras.

López, M. (1988). Hacia una reorientación en la psicología social: después de la crisis. In M. López & R. Zúñiga (Eds.), *Perspectivas críticas en la psicología social* (pp. 383–426). Río Piedras: Editorial de la Universidad de Puerto Rico.

Maldonado, N. (2007). On the coloniality of being. *Cultural Studies, 21*(2), 249–270.

Martín-Baró, I. (1989). *Sistema grupo y poder*. San Salvador: UCA editores.

Martinez Diaz A. (2013). *Autogestion comunitaria*. Retrieved from http://academic.uprm.edu/amartinez/HTMLobj-61/AUTOGESTI_N_COMUNITARIA.doc

Mignolo, W. (2008). La opcion de-colonia:desprendimiento y apertura. Un manifiesto y un caso. *Tabula Rasa, 8*, 243–281.

Mintz, S. (1984). Finding Taso, discovering myself. *Dados, Revista de Ciencias Sociales, 27*(1), 45–58.

Miranda, D. (1995). Women in social movements: Some experiences and issues. In H. Figueroa, M. Lopez, & M. Roman (Eds.), *La Bella Indifferance*. Rió Piedras: Publicaciones Puertorriqueñas.

Miranda, D., De Leon, Y., Franco, M., & Quiñónez, D. (1991). *Community movements in Puerto Rico: It's development and future*. San Jose, Costa Rica: XXIII Interamerican Psychology Congress. Society of Interamerican Psychology.

Miranda, D., Pizarro, D., & Santos Torres, N. (2013). Community movements and the construction of critical citizenship from a decolonizing methodology (translated). In C. P. Echandia, A. D. Gomez, & P. Vamarro (Eds.), *Methodological approaches to political subjectivities: Latinamerican debate (translated)*. Bogota, Colombia: Universidad Distrital Francisco Jose de Caldas: Clacso.

Najenson, J. L. (1981). Democidio: un hijo legitimo del genocidio y destrucción cultural. *Revista de la Universidad Autonoma del Estado de Mexico, 11*, 11–21.

Phillips, N. L., Adams, G., & Salter, P. S. (2015). Beyond adaptation: Decolonizing approaches to coping with oppression. *Journal of Social and Political Psychology, 3*(1), 365–338.

Pizarro, D. (2013). Construcción de ciudadanía desde las luchas comunitarias en Puerto Rico: dimensión política transformadora y la cuestión social [Building citizenship from community struggles in Puerto Rico: political and social transformative dimensión]. In: R. Seda(org) Trabajo comunitario y descolonización [Community endeavor and decolonization]. p. 33-81, Ediciones Callejón.

Povinelli, E. A. (2011). *Economies of abandonment, social belonging and endurance in Late Liberalism*. Durham, NC: Duke University Press.
Red CIMAS. (2015). *Metodologia Participativas, socio praxis para la creatividad social*. Madrid: Dextra Editorial.
Reichman, J., & Fernández Buey, F. (1994). *Redes que dan libertad: introducción a los nuevos movimientos sociales* [Networks that promote liberty: introduction to the new social movements]. Ediciones Paidós.
Retamozo, M. (2006). *El método como postura: apuntes sobre la conformación de la subjetividad epistémica y notas metodológicas sobre la construcción del objeto de estudio*. Proyecto Papime (DGAPA) en 308004. Innovación de métodos, estrategias, materiales didácticos para la enseñanza de la metodología de la investigación.
Reyes Cruz, M., & Sonn, C. (2011). Decolonizing culture in community psychology: Reflections from critical social science. *Journal of Community Psychology, 47*(1–2), 203–214.
Seda, R. (2012). Introducción. En: R. Seda (org) Trabajo comunitario y descolonización [Community work and decolonization]p. 275 – 293. Ediciones Callejón.
Serrano-García, I., & López-Sánchez, G. (1991). Community Interventions in Puerto Rico: The Impact of Social-Community Psychology. *Applied Psychology, 40*(2), 201–218. https://doi.org/10.1111/j.1464-0597.1991.tb01368.x
Sotero, M. M. (2006). A conceptual model of historical trauma: Implications for public health practice and research. *Journal of Health Disparities Research and Practice, 1*(1), 93–108.
Stamm, B. H., Stamm, I. V., Hudnall, A., & Higson-Smith, C. (2014). Considering a theory of cultural trauma and loss. *Journal of Loss and Trauma, 9*, 89–111.
Teo, T. (2015). Critical psychology: A geography of intellectual engagement and resistance. *American Psychologist, 70*(3), 243–254.
Trías Monge, J. (1999). *Puerto Rico: The trials of the oldest colony in the world*. Yale University Press.
Tuhiwai Smith, L. (2012). *Decolonizing methodologies: Research and indigenous people*. London: Zed Books.

Promoting Epistemic Justice: Community Arts, Identity and Belonging Among African Diaspora in Australia

Christopher C. Sonn, Alison M. Baker, and Rama P. Agung-Igusti

This chapter draws on community-engaged research with different collectives made up of people from diverse ethnic and migrant backgrounds, who are negotiating the complex processes of creating social identities and opportunities for belonging and community within a broader local and global context of racialised power relations in Australia. The research is part of a broader initiative called the Community, Identity, Displacement Research Network made up of scholar-activists who seek to bridge the divide between university and community settings through stronger collaborative and community-engaged participatory research aimed at documenting the deleterious psychosocial effects of various forms of symbolic and structural violence such as racism and sexism and creating possibilities for communality, empowerment, and self-determination. In our contexts in Australia, we have examined race and whiteness in everyday settings and institutions, with a focus on how privilege and disadvantage is produced in and through narratives, discourses, and material practices; and the differential implications for belonging, identity, and intergroup relations for people who are othered and excluded through racialisation (e.g., Ali & Sonn, 2010; Green, Sonn, & Matsebula, 2007; Sonn, Ivey, Baker, & Meyer, 2017). These questions have their roots in our own biographies, across our differences and similarities as migrants, and gendered and raced subjects, as we seek to understand and disrupt oppression and injustice. Christopher is an

C. C. Sonn (✉)
Victoria University, Institute of Health and Sport, Melbourne, VIC, Australia

University of the Witwatersrand, Johannesburg, South Africa
e-mail: Christopher.sonn@vu.edu.au

R. P. Agung-Igusti
Victoria University, Institute of Health and Sport, Melbourne, VIC, Australia

A. M. Baker
Victoria University, Institute of Sustainable Industries and Liveable Cities, Melbourne, VIC, Australia

G. Stevens, C. C. Sonn (eds.), *Decoloniality and Epistemic Justice in Contemporary Community Psychology*, Community Psychology,
https://doi.org/10.1007/978-3-030-72220-3_8

immigrant to Australia from South Africa where he lived in what was a classified "coloured" community during the abhorrent Apartheid Era. In Australia he was hailed into a different racial hierarchy, labelled black, and positioned differently in relation to Indigenous blackness, as well as the dominant cultural 'Anglo-Australian' whiteness. Alison is a white settler Canadian and Australian who spent considerable time living in the southern states in the USA and later in El Salvador. She draws on her experiences of migration across these different cultural contexts, with ongoing attention to the ways in which whiteness and white privilege has shaped her own and other's realities. Whilst Rama was born in Australia, he is connected to various migrant histories through his Balinese born father and Austrian born grandmother, and the early anglo-celtic settler heritage of his grandfather. As a new researcher he drew on experiences of his own identity to examine the ways multiracial identities are constructed, positioned and negotiated across contexts for the children of migrants. As a group, we continue to wrestle with the complex, and fraught nature of working together across divides within our group as well as in our partnerships. Given these nuances of our histories and the dynamics within communities, we seek to better understand how racism and sexism get under our skins, looking particularly at the everyday and purposely created settings, (i.e., our community of learning and communities of resistance) that provide people with resources, support and opportunities to remain strong, survive, and experience belonging and communality.

An important element of our efforts has included responding to calls from scholars and activists alike for renewed and reinvigorated epistemologies, methodologies and practices that can contribute to emancipatory aspirations of critical community and liberation psychology (e.g., Dutta, Sonn, & Lykes, 2016; Evans, Duckett, Lawthom, & Kivell, 2017; Fine, 2018; Teo, 2015; Watkins & Shulman, 2008). These calls are not new, but have been given greater impetus by various developments across the globe captured in the growing critical writing on liberation and emancipatory praxis and the decolonial turn (e.g., Adams, Dobles, Gómez, Kurtiş, & Molina, 2015; Adams, Ordóñez, Kurtiş, Molina, & Dobles, 2017; Bhatia, 2007, 2017; Mignolo, 2009; Seedat & Suffla, 2018). We share in this chapter our efforts towards creating inclusive knowledge, a form of critical iterative and generative praxis, borne from our desire for and commitment to dialogical and ethically engaged work that affirms and strengthens people and communities rather than diminish them (Nelson & Prilleltensky, 2010; Sonn & Baker, 2016). We are differently positioned migrants because of various factors including the ideologies of race and white supremacy that shape our experiences and everyday realities. However, among other aspects, we have in common that we are migrants/settlers on Indigenous country, constructing belonging, theorising and developing praxis from this positioning in post-colonizing Australia (Moreton-Robinson, 2003). Indigenous scholar, Aileen Moreton-Robinson (2003) highlighted the different ontological relationship to land that shapes Indigenous belonging, as well as the history and continuity of dispossession that is reflected in pervasive structural inequality and differential access to social, political, and economic power and resources with attendant implications for belonging, community, and wellbeing for Indigenous people and different diasporas. This positioning has become more central in our research and writing as we personally and collectively wrestle with and try to deconstruct whiteness as well as centre the lived experiences and knowledges of various communities whose ways of being, knowing, and doing have been silenced, denied, and neglected (Smith, 2012).

We draw from two recent projects with different collectives constituted by members from the African diaspora who are constructing lives in Australia. These collectives have different goals and aspirations and have been developing their praxis over time with a focus on utilising community arts practices in pursuit of self-determination and self-representation, and to tell their/our stories to offer alternatives to pervasive racialised scripts that constrain the subjectivities and possibilities of racialised/ethnicised people as culturally incompatible others. We navigate different relationships and roles with these groups, such as research collaborators, as we try to configure new ways of working at the university-community interface, vigilant to the dynamics of oppression in research and practice in our home discipline (e.g., Sonn & Green, 2006). Efforts to work in collaborative and horizontal ways have meant learning with and from these groups about the coloniality of power, about their/our lives shaped by racialisation, whiteness, and sexism, the demands of cultural change and identity negotiation in diaspora; and active processes of imagining and creating opportunities for themselves, their communities, and beyond those communities, spaces within which they can metabolise everyday violence and create new narratives for hope and possibilities for belonging.

Race, Whiteness and African Communities in Australia

In Australia, race has been central to formation of the nation since invasion, which was justified through the doctrine of *Terra Nullius,* to the later post-invasion immigration policies (Hage, 1998). In 1788, the British arrived on the Eastern side of Australia in Gadigal territory, a first nations clan belonging to the Eora nation located where Sydney now lies (Heiss & Gibson, n.d.). The land was claimed under the international law of *Terra Nullius*, which posited that land that belonged to no one was free to be claimed by whomever discovered it first. Thus, began the dispossession and violent colonisation of Australia's Indigenous people and the rewriting of Australia's story as both white and British. Indigenous scholar Aileen Moreton-Robinson (2015) speaks of the white possession of Australia, stating that:

> [b]elonging to the new nation… was racialized and inextricably tied to the accumulation of capital and the social worth, authority, and ownership that this conferred. The Indigenous were excluded from this condition of belonging… The white body was the norm and measure for identifying who could belong. (p. 7)

White normativity was reinforced through state institutions and a range of policies known as the "White Australia Policy". These policies effectively barred non-European migration and privileged English migrants, further constructing Australia as a white nation. Whilst, these policies were eventually dismantled in the 1970s in favour of multiculturalism and its economic benefits, belonging was and is still constructed against a normative whiteness reproduced through institutions and legislation that continue to decentre Indigenous sovereignty and racialize non-white migrants (Stratton, 2011).

Different scholars have engaged with the concept of whiteness to bring attention to the reproduction of normativity, dominance, and privilege in relation to Indigenous and migrant communities (Frankenberg, 1993; Green et al., 2007). Hage (1998) noted that whiteness in Australia places white people in dominant cultural positions and grants them unfair privilege, which is not always visible to them. It is recognised by scholars that whiteness is multifaceted, situational and produced and reproduced within specific sets of power relations, and that there is no single narrative of whiteness. Nicolacopoulos and Vassilacopoulos (2004) note that "whiteness is historically and socially constructed through the processes that at once position Indigenous peoples as non-Australian and designated migrant groups as what we might call 'perpetual foreigners' within the Australian state ..." (p. 32). Several researchers have examined whiteness in relation to immigration and ethnic community formation in Australia. These researchers highlight the various ways in which different ethnicised communities are othered and positioned as other against an unexamined or absent standard (e.g., Ali & Sonn, 2010; Tascón, 2008; Zevallos, 2008).

Migrants from the African continent most recently have been subject to racialising discourses and denied belonging. Whilst initial migrants consisted of many white South Africans and Zimbabweans, the 1990s saw an increase of refugees from countries in the Horn of Africa and West Africa, and 2001 onwards saw increases from sub-Saharan Africa. By 2006 African born migrants accounted for 5.4% of Australia's overseas born population (Jakubowicz, 2010). Racism and racialisation has been a significant issue for many of these communities. In a national survey mapping social cohesion it was found that 60–77% of migrants from African countries such as Ethiopia, Kenya, Zimbabwe and South Sudan reported having experienced discrimination, with many also reporting that this was linked to the colour of their skin (Markus, 2016). This was also identified in a report by the Australian Human Rights Commission (2010) that found that "the vast majority of African Australians who participated…said that having a 'visibly different appearance' did impact upon their everyday experiences" (p. 8). Whilst this indicates examples of everyday experiences of racism and discrimination, powerful racialising discourses within the media and politics (Baak, 2018; Wahlquist, 2018; Windle, 2008) uncover how those of African heritage are constructed as criminal, culturally incompatible and unable to be integrated (Baak, 2011). These processes of racialisation produce injustice and position the African diaspora as strangers thereby curtailing their capacity to unconditionally participate in everyday settings and institutions, with attendant consequences for their subjectivities, belonging, health, and wellbeing. Importantly, these communities are not passive victims to dynamics of oppression and exclusion, they find ways to resist, survive, and protect themselves and create opportunities for meaningful engagement, new subject positions, and form alliances with other communities in the pursuit of justice. With this background, we now turn to outlining the theoretical and conceptual coordinates that we have been shaping as part of an ongoing effort to conduct community-engaged research from a liberation standpoint in our effort to create inclusive knowledge.

Displacement, Diaspora and Belonging

There are many approaches to understanding the challenges, tensions, violence, opportunities and radical possibilities provoked by various forms of voluntary and involuntary displacement and the subsequent processes of acculturation, settlement, and belonging. The predominant approach in psychology is based in Berry's (2006, 2011) work on acculturation psychology, which attends to the various changes that occur between individuals and groups as a consequence of ongoing interactions. While this has been a very productive area of work, different scholars have raised questions about the underlying assumptions about identity and culture that underpin acculturation work, presumptions in particular about the desire for assimilation and the stability of identities and categories of group membership. Bhatia and Ram (2001; Bhatia, 2017) maintained that static essentialist understandings of culture and identity are limited, because it is relying on colonial categories derived from Western, Euro-American conceptions that is the basis of an absent standard of comparison. Informed by post-colonial scholarship and narrative psychology, these authors suggest that stories and narratives are vital to understand how people make meaning of their lives and for making visible dynamics of power that is achieved through symbolic violence and expressed in ideologies, values, beliefs and representations that maintain oppression and injustice (Bell, 2010; hooks, 1990; Rappaport, 2000; Smith, 2012; Sonn, Stevens, & Duncan, 2013).

Stuart Hall's (1990) writing, among others (e.g., Anzaldúa, 1987/2007; Bhatia, 2016; Said, 2001), has also been instructive for understanding the complexity of identity for diasporic communities. For Hall, a "Black" immigrant from Jamaica to the UK, cultural identity is not static, it is fluid, contested, and situational, it is a matter of becoming and of being. It is produced in and through narratives of both the past and present, a set of positions, within broader arrangements of power through which people make sense of themselves and their communities. As Hall states, cultural identity is "something—not a mere trick of the imagination. It has histories—and histories have their real, material and symbolic effects. The past continues to speak to us. ... It is always constructed through memory, fantasy, narrative and myth. Cultural identities are the points of identification or suture, which are made, within the discourses of history and culture. Not an essence but a *positioning*" (p. 226).

Ahmed (2000), writing in the context of transnationalism, contends that immigrants have to negotiate many layers of being made strange through being constructed as a different other. And the new rise of transnational feminist scholars insist on theorizing how structural intersectionalities shape individual and collective subjectivities, at the nexus of race/ethnicity, culture, gender, sexuality, immigration status, age, and (dis)ability (Anzaldúa, 1987/2007; Collins, 2000, 2002; Yuval-Davis, 2006). Both hyper visible and (in)visible, migrants are not simply unrecognisable, they are often deeply known through social and cultural processes that rely on binaries of inclusion-exclusion that are encoded in and through bodies, accents, skin colour, physiognomy, dress and so on. These processes function to maintain self-other positions that produce and reproduce various unequal social structural arrangements. Feminist scholarship has highlighted the intersections of multiple identities and oppressions within power arrangements as well as the agentic self, capable of resistance and self-determination (e.g., Anzaldúa, 1987/2007; Collins, 2000, 2002; Lugones, 1992). Maria Lugones (1992), for example, built on Anzaldua's notion of mestizaje consciousness to develop the notion of "active subjectivity" to show how displaced and oppressed groups resist in various ways to protect identities and navigate the social world both "outside" the group, and within. This critical feminist scholarship has been central to new epistemologies anchored in lived experiences, relationality, ethics of care, and attuned to the multiple social locations, borders, and resources that people negotiate as they construct subjectivities and ways of belonging.

Informed by this critical scholarship, we now turn to two participatory community arts projects that used storytelling as a pivotal methodology for social change. These projects are sites of political and aesthetic resistance enacted through the creation of settings in which people are able to interrogate and produce narratives about place, community, and belonging (Sonn & Baker, 2016; Sonn, Smith & Meyer, 2015). As Segalo, Manoff, and Fine (2015) noted, decolonisation requires the creation of "spaces where bodies, affect, rage, struggle, imagination can breathe" (p. 19). Our focus is on understanding how community arts spaces promote imagination, the functions of narrating lived experiences, and the implications of these on self, community, and solidarities across social and racial groups.

Community Arts, Epistemic Justice, and African Diaspora

In this section, we describe the work produced in partnerships with community-based creative collectives, cohealth Arts Generator (cAG) and New Change, both located in Melbourne, Australia, who have various agendas including empowerment, health promotion, education and social justice education. We focus on the ways in which these groups are destabilising the coloniality of power and being, and creating opportunities for connection, resistance, and producing public pedagogies through community arts practice.

cohealth Arts Generator

cAG is a participatory arts "space" that mobilise arts-based programs and opportunities to increase agency and improve wellbeing for young people from diverse social and cultural backgrounds. At the forefront of this participatory model is the privileging of lived experience, the creation of spaces to speak up and push back, development of critical literacies within the community, and the collaborative creation of new knowledges. Liss Gabb (2017), the program coordinator at the Arts Generator, conceptualises the work as producing contact zones (Pratt, 1991), "where public pedagogy meets race literacy and arts practice". cAG, have over the years, developed models of practice based on mobilising encounters with difference as a catalyst for meaningful transformation; through its participatory and community arts programs, it has challenged structural systems of oppression within Australia, by centering there lived experiences and community stories and decentring whiteness in everyday and institutional contexts. The most recent project produced by cAG, and on which the authors worked as research collaborators, to document, codesign research, evaluate, and contribute through participation in reflective and planning sessions, is the focus of this chapter. The key questions that were developed for research and evaluation set out to document the shared narratives of the African diaspora in Australia, and the experiences of both the performers and audiences who witnessed the performance of the diverse stories. Three aspects of how community drama produced liberatory effects at individual, group and intergroup levels are outlined.

The project was born about of the desire of young African Australians from South Sudan, Liberia, Burundi, Uganda, Ghana, Ethiopia and Tanzania, some who have been working as creatives with cAG for over 8 years. Using community drama the collective wanted to insert alternative representations centred on their experiences into a public discourse dominated by essentialised, restrictive and often damaging constructions of Africans in Australia. For authentic representation of themselves, the artists needed to have complete autonomy of their stories, embodying processes of self-determination and self-representation within the creative project. To do this they consciously located themselves and their relationship with the coloniality of Australian society, and re-centred the African narrative. They intentionally de-centred aspects of whiteness, Eurocentric ideas and processes, and tried to minimise using "mainstream" processes as peripheral tools, resources and support. The collective co-created the production that explored identity, dispossession and cultural conflict within the African Australian experience. The performance combined poetry, music, dance, theatre and projection through a series of individual and ensemble pieces that interpret stories of the African diaspora in Melbourne. In terms of process, the group sought to enact horizontal structures such as having no single director or form of hierarchy in the show or entire process, that is, each person played an equal role in the development (Sonn, Agung-Igusti, & Komba, 2018). The process, like the content, is an act of resistance, to what they viewed as Eurocentric and hierarchical structures and processes.

The stories that were told through interviews and that were at the centre of the performance captured the symbolic and epistemic violence in everyday life for many members of the diaspora, yet also moved beyond a homogenous and monolithic single story, capturing the multitudinous and nuanced complexities of lived experience and subjectivity. These stories made visible the racialised power relations, which constrain belonging within the Australian context, but also the acute awareness of difference and sameness and the contingent and contested nature of diasporic belonging. One of the performers interviewed expressed the "tricky-ness" and layered complexity of African identity making in the diaspora in the following way:

> … in the Australian context, it's been really tricky, because I'm identified as African, and I can identify as African. I have absolutely no problem with it. But, I think most of the time when people identify you with it they do place a limit and they have a narrow scope with which they look at it from 'cause when I identify as African, I know it's completely different to when my dad says he identifies as African, and then I also identify very strongly as Black, globally 'cause I'm placed as Black. But then in Australia it's quite tricky, 'cause like I'm the type of Black that's got to recognise that I am on stolen land, and that I am here potentially as a coloniser if I'm not fully aware of my footprint and what it is I'm doing, so…that's quite different to being Black elsewhere. (wāni, Bashi Tribe from Walungu)

This excerpt captures some of the complexity of identity negotiation for diasporic youth. They navigate multiple narratives through which individuals position themselves and are positioned as well as the transnational, and local narratives of Africans, blackness, and how these are expressed in relation to Indigenous blackness. The stories told by these young performers, disrupted singular dominant group representations of African identities offering an understanding of a language to comprehend their complex, fluid, intersectional selves. The group challenged the homogenising and demonising effects of racialised oppression, and instead forge identities, as a process of becoming, out of multiple presences as diasporic subjects within the discourses of history and culture (Bhatia, 2017; Hall, 2000).

For many of the group, this project and experiences within the collective was part of an ongoing process of community-making, of constructing a home-place (Collins, 2000) within which they could share, provide mutual support, laugh, and wrestle with the deep psychopolitical dynamics of self-determination, away from the white gaze. In line with Segalo et al. (2015), creating spaces are an important aspect of the process of doing personal liberation work, of making meaning of decolonial discourse and practice, individually and collectively, and forging speaking positions robustly secured in the ongoing process of recovering and rearticulating their cultural inheritance in Australia. The consciousness raising process was not easy; it entailed charting into unfamiliar territory and carving out a "new pathway" on terms set by them. Another group member reflected:

> We were carving out a new pathway. That was hard, it was always going to be difficult but we knew it had to be done, and on our own terms. It was rewarding because of the richness and rawness in how we were all able to share our own stories, our fears, trauma and aspirations as authentically as possible without it being played around with too much. This was important but it couldn't be done without trust, giving up space for others to be heard and letting go. (Geskeva, Tanzania and Comoros Islands)

Researching, retrieving, and reconnecting with community and diaspora cultural knowledge can be emotionally fraught, especially when situated within oppressive histories and tethered to lived experiences of the present. This accentuates the importance of supportive and safe spaces for the process of retrieving cultural resources and community histories in the liberation process. Drawing on the past and the present to imagine the future is captured in the following Sudanese proverb used within the performance: "We desire to bequeath two things to our children; the first one is roots, the other one is wings". The search for and use of these proverbs are symbolic actions of reclamation that provide mediational tools for making sense of lives, aspirations, sacrifices and roles. While the main focus was to retrieve and recentre African wisdom, whiteness was also decentred by the same process.

The AMKA project focussed on reaching the African diaspora, inevitably it also reached the various communities that make up multicultural Melbourne. In this instance, this community theatre was a form of public pedagogy that could produce transformative witnessing (Sajnani, 2010), and foster consciousness or wide-awareness (Greene, 1995). This witnessing can be facilitated through a reduction of aesthetic distance, the protective space between audience and performer that mediates immersion between the world of the performance or the audience's own reality (Wood, 2017). The psychological distance can be closed as individuals identify their own experiences opening up the possibility for aesthetic intimacy (Greene, 1995). Analysis of audience survey data showed that for many audience members from the African diaspora encountering the stories via theatre resonated with them, it affirmed and validated their own experiences and the complexities of diasporic belonging. Many audience members from other diasporas were also able to identify threads of their own community narratives reflected in the performance, especially the processes of racialisation, othering and negotiating belonging. For people who described themselves as white or Anglo-Australian, there were different effects of witnessing. Responses indicated increased awareness of their own prejudices and privilege, and lack of knowledge about the experiences of Africans in Australia. As one audience member wrote:

> The themes and dilemmas presented were plenty. The dilemma of living in the margins was tackled from the onset. The sense of anger, passion, hunger and hopelessness burst on to the stage. "This is an angry poem" echoed the frustration which accompanies prolonged patience with injustice and neglect. The challenge of "privilege" was enough to make audience members stir in their seats. If aware of it, it was strong enough to evoke a discomfort that accompanies the shame of knowing and not acting. If unaware, it evokes a shame of learning for the first time. For others, it will have been dismissed as banter between mates, but this is the nature of prejudice and art; it's complex.

This excerpt captures the different resonances for audience members. The witnessing triggered reflexivity in relation to the diversity of white subjectivities and the various affective responses to the stories of "injustice and neglect". Discomfort and shame produced by both knowing and not acting and not knowing, epistemological ignorance (Sullivan, 2006), are important to interrogate in the process of charting pathways, rooted in critical reflexivity, to mutual accompaniment and ways to decolonize whiteness (Watkins, this volume).

New Change

New Change is an arts collective of ten young African women aged 18–23 of primarily South Sudanese background in Melbourne's western suburbs. The group's aim is to empower, motivate and educate young women to develop self-confidence, leadership and advocacy skills. The group uses (collective) spoken word poetry, dance and film to explore topics and issues that are relevant to their lives, with a focus on three core elements: creating collective arts works to be used for social change; engaging in public outreach (visiting schools, running workshops and delivering public talks/performances); and activism and advocacy. The issues they focus on and the stories they create through these arts mediums reflect many experiences linked to their settlement journeys in Australia as well as aspects of their daily lives as young people. The members bring to work knowledge from other community engagement, in addition to employment/tertiary studies, including youth work, community development, psychology, health sciences and law.

Originally developed as a program through a local council in 2014 the group tackled topics of importance, including sexual health and misrepresentation of the African Australian community in the media. During 2016, in response to the negative, racialised news coverage of African young people at a community festival, the group sought to challenge this ongoing vilification by media and politicians alike. This project allowed New Change, still a program in Youth Services of a Local Government, to shift to an arts-for-social-change framework, working with independent artists of African heritage to create "Breaking News" a film clip that featured spoken word poetry and dance (see: https://youtu.be/hcEoIucI3tg). In the video, young women rehearse a collective spoken word poem which lays out "a battle for truth" in which they "fight with their words" to challenge negative stereotyping of young African Australians and develop a strong counter-narrative about the South Sudanese community. Yet while the creation of the video provided important opportunities for the young women to "speak their truth" and gain new skills in art-making and public speaking, it wasn't without tensions. As one member reflects on this experience, she notes that there was silencing occurring:

> …even the council there was a lot of conflict going on…within the program. That was a big issue because we had to be careful, we couldn't speak the truth full on…some words had to be edited… for the council to be like "yes can use that…or you can't use that because you are targeting the white community"… We weren't trying to create conflict we were just trying to let the community know [about what we were facing].

In 2017, the group ceased being a council run program and shifted to become an independent arts collective in order to ensure uncontested freedom of expression, but also to be able to focus on transitioning into a "for us by us" entity. The council had developed a relationship with the group of young women and has continued to support them through a local artists residency program, providing space, start-up funds and casual employment for two members. This transition also resulted in collaboration with the university and a move into a participatory action research project using arts to continue to explore relevant topics, as well as develop capacities to move into a collective led solely by the young women. The focus of the research has been to better theorise how young women in New Change use arts to explore identity, culture and voice for empowerment in a context that has highlighted the vilification, misrepresentation and experiences of racism (Baak, 2011; Nunn, 2010; Run, 2013; Windle, 2008). These contexts have also rendered young South Sudanese women "invisible" not only in Australian society but to some extent within their own community as a result of particular age-gender hierarchies (Grabska, 2010; Harris, 2010).

In 2017, New Change members spoke or performed at a dozen schools, at conferences, on radio and media outlets, an international conference and to a number of government officials, including at the Federal Parliamentary Inquiry into migration and settlement outcomes. The group was continuously being invited to tell their stories, especially those experiences relating to the so-called "Apex gang", a "hot topic" in the public realm. Many New Change members became tired of speaking back to "whiteness", the persistent othering, with little room to expand and project their voices and their complex personhoods. The balancing act between "educating" the public and ensuring their not being cornered into the "single story" has been an ongoing challenge as noted by a group member:

> They don't want to get to know young people, they don't want to get to know their real lives… Everything that we were talking about—it could be [they ask you] *what are your hobbies?*—oh I like dancing, I like drawing, I like this… [and they would say] *Oh, you said you like dancing, when you dance do you express yourself through emotion and when you are emotional is it because of all the racism going on? Is it because of how you came to Australia?* Everything would be tied back to racism, arriving to Australia, school, but just all negative. You will just have to advocate, advocate, advocate. Speak the truth. Truth that they don't care about. You're trying hard to let them know, we are good.

Group members were set on arts projects that reflected their daily lives, who they are individually, their dreams, aspirations, and their hardships, and some of the issues they believed needed attention within their own families and community. The project *Understand Me* grew from this desire. Many members of the group had been engaging in ongoing conversations about feelings of isolation that were considered taboo in their cultural community. Silence across generations and between friends often compounded such feelings and the group was intent on starting a conversation within the African community in Australia. They felt the weight of the slow violence of racialization in broader society and various settings such as school and sport clubs (Baker-Lewton, Sonn, Vincent, & Curnow, 2017), but also the weight of family expectations, deciding their futures and figuring out who they are. Thus, the project was not developed solely from their experiences linked to displacement, migration and being "in-between" cultures, but the everyday feelings of isolation and sadness that many young people face when carving out their lives.

Name Poems

The group embarked on their first film, enlisting the help of a local participatory film company to train them in all aspects of filmmaking. Over the course of a few months and six intensive sessions, the group learned the basics and collectively conceptualized the idea and script and filmed the clip. Five group members participated in editing sessions and the film was edited into a 3-min piece that features a series of poetic letters written to their parents, "from" (the perspective) of their parents, to friends and to the world (see: https://youtu.be/C7HRvRCvxeE). The film clip was launched at a local community-university arts space, with the primary audience being young people and parents from the African community. To create a dialogue the clip inspired a sound and a multi-channel video installation that featured additional material including: name poems of each New Change member, translations of the poems to and from parents into Dinka and Arabic, family photographs and scenes not included in the shorter clip.

Name poems became a significant element of the installation, but also subsequent performances and public speaking events (see: https://soundcloud.com/new-change-399717311/name-poems). These short poems were played back or performed before the start of each event as a way of opening and claiming space, about positioning themselves and the listener. As with the two examples below, each poem involved each person's full name, a poetic reflection about the meaning of their name, the line "I remember the first time I found my voice…," a line that detailed what it felt like or how they found their voice and three traits or qualities about themselves.

> Make the change don't just speak it.
> Nyanwel Deng Malual Aweng Machol
> One thousand lessons in one name
> I am Awel.
> I remember the first time I found my voice
> I woke to tell my truth.
>
> – Powerful. Wise. Courageous. (Entire group)

As part of the processes of intergroup relations and migrant belonging, immigrants sometimes anglicize their names in public and work settings to fit in or pass or make it easier for members of the dominant culture. The name poems, are a decolonial action that entailed re-asserting identity and culture, and bringing meaning to the forefront, something that is often ignored in Australian society. Many young women said they "lost" their names in coming to Australia because teachers and others would pronounce them incorrectly. This shedding of their name is in response to microaggressions, and the pressures brought by expectations that integration entails embracing whiteness. Using their languages asserts and revives cultural knowledge that has become marginal and "un-Australian", it positions young women as the knowledge holder, even making the listener uncomfortable because they cannot understand and are therefore not included.

Conclusion

In this chapter we presented work with two arts collectives constituted by majority members from the African diaspora to understand how they are mobilising creative and cultural approaches and knowledge to promote to self-determination and tackle symbolic violence. As members of a research collective, we have been acutely aware of our disciplines complicity in producing epistemic violence through colonising knowledge production practices. We have joined a long list of critical scholars who are generating alternative, participatory, and empowering ways to produce research and contribute to liberation efforts recognizing this as ongoing iterative and generative work (Fine, 2012, 2018; Gergen & Gergen, 2010, 2011; Montero, 2009; Teo, 2015). We highlighted the stories of different young creatives who are negotiating everyday and institutional racism and making sense of displacement, intercultural and intergenerational experiences within a local-global context at the intersections of gender, race/ethnicity, and generational status. Their stories convey the violence of a racism rooted in colonialism and perpetuated through the coloniality of power (Quijano, 2000), which is expressed through epistemic and symbolic violence in media and everyday institutions. The stories shared across the projects provide glimpses of the complex and layered ways in which racism is experienced ranging from not being able to speak their truth, surveillance, and being locked into predetermined colonial stereotypes of Africa rooted in Eurocentrism and whiteness (Cesaire, 2000; Fanon, 1967). The stories also provide insight into the dynamics of resistance involved in negotiating identities and belonging for diasporic youth. In particular, the cases show how the young people carved out social spaces, sometimes away from the controlling white spaces that censors their truths. In the spaces that they create group members share connections, experiences, and create communality—their spaces are home places (Collins, 2000). In those spaces they express and articulate social and psychological experiences and develop a language for their realities. As "active subjects" (Lugones, 1992) the young people deconstruct and reconstruct different experiences, memories, and knowledge as they negotiate being and belonging and express their diverse social identities as African young people in Australia. hooks (1990) commented that "fundamental to the process of decentering the oppressive other and claiming our right to subjectivity is the insistence that we must determine how we will be and not rely on colonising responses to our own legitimacy" (p. 22).

Both groups have used art and creative expression as a key modality for their political work that is aimed at naming their realities and claiming ways to belong, crafting stories rooted in their lived experiences which they have shared with the broader public. Their stories were tied to African community histories that are so often invisible or omitted reduced to stereotypes of poverty, trauma, and criminality. Resources retrieved from the rich diversity of African cultures and histories were infused into their arts products that were projected into the public sphere as an act of community affirmation and of critical counter storytelling (Bell, 2010). Arts modalities, such as community drama and popular theatre, were used not only as outputs or for translation, but as meaning-making and knowledge creation, performance and transformation (Boal, 1998; Freire, 1972; Gergen & Gergen, 2011; Greene, 1995). In this case, the young people created and performed potent narratives that transverse the space between performer and witness, provoking a reflexivity that fosters critical consciousness or contemplation of one's own complicity in sustaining the oppressive structures of whiteness. Arts are often deeply affective and reflexive, and through these qualities art becomes a force for transformation and resignification, not merely an aesthetic artefact but a political act by connecting the tendrils of oppression in personal lives to broader collective narratives, histories, and structures, and imagining alternative futures.

As we conclude, we recognize that we are part of and implicated in Australia's history, it is not in the past, and we seek to respond and contribute to creating inclusive knowledge and receptive communities. Thus, we return to our goals of wanting to contribute to critical community psychology through "research that celebrates survival and resilience and that revitalizes language, arts, and cultural practices" (Watkins & Shulman, 2008, p. 276). In doing this we continue to construct a community of learning within the neoliberalising university. In this collective we are interrogating our own positioning and biographies, while also wrestling with the tensions and challenges arising out of our commitment to recover historical memory and to interrogate racism and whiteness together from our diverse entangled positions and consciousness within the matrix of power. This work is by no means easy involving significant emotional labour as we go about navigating the various borders that separate and connect us. However, through our partnerships within and outside the university we have found resonant voices, a groundswell of people seeking new ways of being, doing, and knowing. With them and colleagues in different countries we want to engage in the critical work of building solidarities and interrogate the dynamics of oppression in research and action in and beyond the university.

References

Adams, G., Dobles, I., Gómez, L. H., Kurtiş, T., & Molina, L. E. (2015). Decolonizing psychological science: Introduction to the special thematic section. *Journal of Social and Political Psychology, 3*(1), 213–238. https://doi.org/10.5964/jspp.v3i1.564

Adams, G., Ordóñez, L., Kurtiş, T., Molina, L., & Dobles, I. (2017). Notes on decolonizing psychology: From one special issue to another. *South African Journal of Psychology, 47*, 531–541. https://doi.org/10.1177/0081246317738173

Ahmed, S. (2000). *Strange encounters*. London, UK: Routledge.

Ali, L., & Sonn, C. C. (2010). Constructing identity as a second-generation Cypriot Turkish in Australia: The multi-hyphenated other. *Culture & Psychology, 16*(3), 416–436. https://doi.org/10.1177/1354067X10361398

Anzaldúa, G. (2007). *Borderlands/La Frontera: The new mestiza* (3rd ed.). San Francisco: Aunt Lute Books. (Original work published 1987).

Australian Human Rights Commission. (2010). *In our own words. African Australians: A review of human rights and social inclusion issues.* Retrieved from https://www.humanrights.gov.au/sites/default/files/content/africanaus/review/in_our_own_words.pdf
Baak, M. (2011). Murder, community talk and belonging: An exploration of Sudanese community responses to murder in Australia. *African Identities, 9*(4), 417–434. https://doi.org/10.1080/14725843.2011.614415
Baak, M. (2018). Racism and othering for South Sudanese heritage students in Australian schools: Is inclusion possible? *International Journal of Inclusive Education, 23*(1), 1–17. https://doi.org/10.1080/13603116.2018.1426052
Baker-Lewton, A., Sonn, C. C., Vincent, D. N., & Curnow, F. (2017). 'I haven't lost hope of reaching out…': Exposing racism in sport by elevating counter narratives. *International Journal of Inclusive Education, 21*(11), 1–16. https://doi.org/10.1080/13603116.2017.1350316
Bell, L. A. (2010). *Storytelling for social justice: Connecting narrative and arts in antiracist teaching.* New York, NY: Taylor & Francis.
Berry, J. W. (2006). Contexts of acculturation. In D. L. Sam & J. W. Berry (Eds.), *The Cambridge handbook of acculturation psychology* (pp. 27–42). New York: Cambridge University Press.
Berry, J. W. (2011). Immigrant acculturation: Psychological and social adaptations. In A. E. Azzi, X. Chryssochoou, B. Klandermans, & B. Simon (Eds.), *Identity and participation in culturally diverse societies: A multidisciplinary perspective* (pp. 279–295). Oxford, UK: Wiley-Blackwell.
Bhatia, S. (2007). *American karma: Race, culture, and identity in the Indian diaspora.* New York: NYU Press.
Bhatia, S. (2016). Social psychology and social justice: Citizenship and migrant identity in the post 9/11 era. In P. L. Hammack Jr. (Ed.), *The Oxford handbook of social psychology and social justice.* https://doi.org/10.1093/oxfordhb/9780199938735.013.17
Bhatia, S. (2017). *Decolonising psychology: Globalisation, social justice and Indian youth identities.* London, UK: Oxford University Press.
Bhatia, S., & Ram, A. (2001). Rethinking 'acculturation' in relation to diasporic cultures and postcolonial identities. *Human Development, 44*(1), 1–18.
Boal, A. (1998). *Legislative theatre: Using performance to make politics.* London, UK: Routledge.
Cesaire, A. (2000). *Discourse on colonialism* (Trans. by J. Pinkman). New York: Monthly Review Press.
Collins, P. H. (2000). Gender, black feminism, and black political economy. *The Annals of the American Academy of Political and Social Science, 568*(1), 41–53.
Collins, P. H. (2002). *Black feminist thought: Knowledge, consciousness and the politics of empowerment* (2nd ed.). New York, NY: Routledge.
Dutta, U., Sonn, C. C., & Lykes, M. B. (2016). Situating and contesting structural violence in community-based research and action. *Community Psychology in Global Perspective, 2*(2), 1–20.
Evans, S. D., Duckett, P., Lawthom, R., & Kivell, N. (2017). Positioning the critical in community psychology. In M. A. Bond, I. Serrano-Garcia, C. B. Keys, & M. Shinn (Eds.), *APA handbooks in psychology. APA handbook of community psychology: Theoretical foundations, core concepts, and emerging challenges* (pp. 107–127). Washington, DC: American Psychological Association.
Fanon, F. (1967). *Black skins, white masks.* New York: Grove Press.
Fine, M. (2012). Troubling calls for evidence: A critical race, class and gender analysis of whose evidence counts. *Feminism & Psychology, 22*(1), 3–19. https://doi.org/10.1177/0959353511435475
Fine, M. (2018). *Just research. Widening the methodological imagination in contentious times.* New York: Teachers College Press.
Frankenberg, R. (1993). *White women, race matters: The social construction of whiteness.* Minneapolis, MN: University of Minnesota Press.
Freire, P. (1972). *Pedagogy of the oppressed.* Ringwood, Australia: Penguin.
Gabb, L. (2017, December). *Contact zone: Where public pedagogy meets race literacy and arts practice.* Keynote presented at the Public Pedagogies Institute Conference.

Gergen, K. J., & Gergen, M. M. (2010). Scanning the landscape of narrative inquiry. *Social and Personality Psychology Compass, 4*(9), 728–735.

Gergen, M. M., & Gergen, K. J. (2011). Performative social science and psychology. *Historical Social Research/Historische Sozialforschung, 36*(4), 291–299. https://doi.org/10.2307/23032295

Grabska, K. (2010). Lost boys, invisible girls: Stories of Sudanese marriages across borders. *Gender, Place & Culture: A Journal of Feminist Geography, 17*(4), 479–497. https://doi.org/10.1080/0966369X.2010.485839

Green, M. J., Sonn, C. C., & Matsebula, J. (2007). Reviewing whiteness: Theory, research possibilities. *South Africa Journal of Psychology, 37*(3), 389–419.

Greene, M. (1995). *Releasing the imagination: Essay on education, the arts, and social change*. San Francisco, CA: Jossey-Bass.

Hage, G. (1998). *White nation: Fantasies of white supremacy in a multicultural society*. Sydney: Pluto Press.

Hall, S. (1990). Cultural identity and diaspora. In J. Rutherford (Ed.), *Identity: Community, culture difference* (pp. 222–237). London: Lawrence and Wishart.

Hall, S. (2000). Who needs identity? In P. du Gay, J. Evans, & P. Redman (Eds.), *Identity: A reader* (pp. 15–30). IDE: Sage.

Harris, A. (2010). Race and refugeity: Ethnocinema as racial pedagogy. *Qualitative Inquiry, 16*(9), 768–777. https://doi.org/10.1177/1077800410374445

Heiss, A., & Gibson, M. (n.d.). *Aboriginal people and place*. Retrieved from http://www.sydney-barani.com.au/sites/aboriginal-people-and-place/

hooks, B. (1990). *Yearning: Race, gender, and cultural politics*. Boston, MA: South End Press.

Jakubowicz, A. (2010). *Australia's migration policies: African dimensions*. Retrieved from https://www.humanrights.gov.au/sites/default/files/content/africanaus/papers/africanaus_paper_jakubowicz.pdf

Lugones, M. (1992). On Borderlands/La Frontera: An interpretive essay. *Hypatia, 7*(4), 31–37.

Markus, A. (2016). *Australians Today: The Australia@2015 Scanlon Foundation Survey*. Retrieved from http://scanlonfoundation.org.au/wp-content/uploads/2016/08/Australians-Today.pdf

Mignolo, W. (2009). Epistemic disobedience, independent thought, and de-colonial freedom. *Theory, Culture, and Society, 26*, 1–23. https://doi.org/10.1177/0263276409349275

Montero, M. (2009). Methods for liberation: Critical consciousness in action. In M. Montero & C. Sonn (Eds.), *Psychology of liberation. Theory and applications* (pp. 73–92). New York: Springer.

Moreton-Robinson, A. (2003). I still call Australia home: Indigenous belonging in a white society. In S. Ahmed, C. Castaneda, A. Fortier, & M. Sheller (Eds.), *Uprootings/regroundings: Questions of home and migration* (pp. 23–40). Oxford, England: Berg.

Moreton-Robinson, A. (2015). *The white possessive: White possession and indigenous sovereignty matters*. Minneapolis, London: University of Minnesota Press.

Nelson, G. B., & Prilleltensky, I. (2010). *Community psychology: In pursuit of liberation and well-being*. Basingstoke, New York: Palgrave Macmillan.

Nicolacopoulos, T., & Vassilacopoulos, G. (2004). Racism, foreigner communities and the onto-pathology of white Australian subjectivity. In A. Moreton-Robinson (Ed.), *Whitening race: Essays in social and cultural racism* (pp. 32–47). Canberra: Aboriginal Studies Press.

Nunn, C. (2010). Spaces to speak: Challenging representations of Sudanese-Australians. *Journal of Intercultural Studies, 31*(2), 183–198. https://doi.org/10.1080/07256861003606366

Pratt, M. L. (1991). Arts of the contact zone. *Profession*, 33–40.

Quijano, A. (2000). Coloniality of power, eurocentrism, and Latin America. *Nepantla: Views from South, 1*(3), 533–580.

Rappaport, J. (2000). Community narratives: Tales of terror and joy. *American Journal of Community Psychology, 28*(1), 1–24. https://doi.org/10.1023/A:1005161528817

Run, P. (2013). Unnecessary encounters: South Sudanese refugees' experiences of racial profiling in Melbourne. *Social Alternatives, 32*(3), 20–25.

Said, E. W. (2001). *Reflections on exile and other literary and cultural essays*. London: Granta.
Sajnani, N. (2010). Mind the gap: Facilitating transformative witnessing amongst audiences. In P. Jones (Ed.), *Drama as therapy* (Vol. 2). London: Routledge.
Seedat, M., & Suffla, S. (2018). Community psychology and its (dis)contents, archival legacies and decolonisation. *South African Journal of Psychology, 47*, 421–431. https://doi.org/10.1177/0081246317741423
Segalo, P., Manoff, E., & Fine, M. (2015). Working with embroideries and counter-maps: Engaging memory and imagination within decolonizing frameworks. *Journal of Social and Political Psychology, 3*(1), 342–364. https://doi.org/10.5964/jspp.v3i1.145
Smith, L. T. (2012). *Decolonizing methodologies: Research and indigenous peoples* (2nd ed.). London, England: Zed Books.
Sonn, C. C., Agung-Igusti, R., & Komba, G. (2018). *Theatre for change: Witness and performer responses to African stories of identity, belonging and community*. Melbourne, VIC: Victoria University.
Sonn, C. C., & Baker, A. M. (2016). Creating inclusive knowledges: Exploring the transformative potential of arts and cultural practice. *International Journal of Inclusive Education, 20*(3), 215–228. https://doi.org/10.1080/13603116.2015.1047663
Sonn, C. C., & Green, M. J. (2006). Disrupting the dynamics of oppression in intercultural research and practice. *Journal of Community & Applied Social Psychology, 16*(5), 337–346.
Sonn, C. C., Ivey, G., Baker, A., & Meyer, K. (2017). Post-apartheid South African immigrants in Australia: Negotiating displacement, identity and belonging. *Qualitative Psychology, 4*(1), 41–54.
Sonn, C. C., Stevens, G., & Duncan, N. (2013). Decolonisation, critical methodologies and why stories matter. In G. Stevens, N. Duncan, & D. Hook (Eds.), *Race, memory and the apartheid archive* (p. 295). London, UK: Palgrave Macmillan.
Sonn C., Smith K., & Meyer K. (2015). Challenging Structural Violence Through Community Drama: Exploring Theatre as Transformative Praxis. In D. Bretherton, S. Law (Eds). Methodologies in Peace Psychology. Peace Psychology Book Series, vol 26. Springer, Cham. https://doi.org/10.1007/978-3-319-18395-4_15
Stratton, J. (2011). The Jackson Jive: Blackface today and the limits of whiteness in Australia. *The Journal of European Associations of Studies on Australia, 2*(2), 22–41.
Sullivan, S. (2006). Race, space, and place. In *Revealing whiteness: The unconscious habits of white privilege* (pp. 143–166). Bloomington, IN: Indiana University Press.
Tascón, S. M. (2008). Narratives of race and nation: Everyday whiteness in Australia. *Social Identities, 14*(2), 253–274. https://doi.org/10.1080/13504630801933688
Teo, T. (2015). Critical psychology: A geography of intellectual engagement and resistance. *American Psychologist, 70*(3), 243–254. https://doi.org/10.1037/a0038727
Wahlquist, C. (2018, January 3). Is Melbourne in the grip of African crime gangs? The facts behind the lurid headlines. *The Guardian*. Retrieved from https://www.theguardian.com/australia-news/2018/jan/03/is-melbourne-in-the-grip-of-african-gangs-the-facts-behind-the-lurid-headlines
Watkins, M., & Shulman, L. (2008). *Toward psychologies of liberation*. New York, NY: Palgrave Macmillan.
Windle, J. (2008). The racialisation of African youth in Australia. *Social Identities, 14*(5), 553–566. https://doi.org/10.1080/13504630802343382
Wood, S. M. (2017). *Witnessing stories: The transformative impact of witnessing performed lived experience within the context of family life education* (Unpublished doctoral dissertation). Concordia University, Montreal, Quebec, Canada.
Yuval-Davis, N. (2006). Belonging and the politics of belonging. *Patterns of Prejudice, 40*(3), 197–214.
Zevallos, Z. (2008). 'You have to be Anglo and not look like me': Identity and belonging among young women of Turkish and Latin American backgrounds in Melbourne, Australia. *Australian Geographer, 39*(1), 21–43. https://doi.org/10.1080/00049180701877410

Ethnic Conflict and Peacebuilding in Northeast India: A Decolonial Perspective

Urmitapa Dutta

Foregrounding "Context:" Ethnic Conflict in Northeast India

Northeast India constitutes the easternmost landmass of the Indian state. A narrow strip of land connects the region to the rest of India, referred to as "mainland India" in popular discourse. "Northeast" is an administrative category coined during British colonial rule, which designated vast portions of the region as "backward" and "excluded" areas—designations that continue to inform contemporary public policy (Baruah, 2005). Consisting of eight[1] states, Northeast India is characterized by extraordinary ethnocultural and linguistic diversity with significant "tribal" populations.[2] The term "tribe" is a colonial construction that conflates diverse groups with mutually unintelligible languages, cultures, and religions to distinguish them from the category of "caste." Both popular and public policy discourses frame tribes from Northeast India as self-contained primitive social formations that are culturally and developmentally inferior (Bal, 2007). Simultaneously, the term also constitutes a political and cultural identity around which demands for self-determination are mobilized. Then there is the state-sponsored category of "non-tribals," a

[1] Historically the region consisted of seven states—Assam, Arunachal Pradesh, Manipur, Mizoram, Meghalaya, Nagaland, and Tripura. The eastern hill state of Sikkim was added to the category more recently.

[2] Basic Statistics of North Eastern Region 2015, North Eastern Council Secretariat, Government of India.

U. Dutta (✉)
Department of Psychology, University of Massachusetts Lowell, Lowell, MA, USA
e-mail: Urmitapa_Dutta@uml.edu

G. Stevens, C. C. Sonn (eds.), *Decoloniality and Epistemic Justice in Contemporary Community Psychology*, Community Psychology,
https://doi.org/10.1007/978-3-030-72220-3_9

monolithic category that includes ethnic communities that are long term residents as well as recent migrants from neighboring countries and other parts of India.[3]

Ethnic conflicts in different parts of Northeast India have varied timelines, some dating back to 70 years. These struggles are mobilized around demands for autonomy and self-determination, both within and in secession from the Indian state. In the Garo Hills region, located in the Northeastern state of Meghalaya (where I was born and raised), the genesis of ethnonationalist conflict, including armed insurgency, can be traced to the early 1990s (Singha & Nayak, 2015). The Garos are the dominant tribal group in Garo Hills with non-tribals constituting the "ethnic others." Armed insurgency here is based on demands for a "Greater Garoland," an autonomous state within India comprising of the present region of Garo Hills and contiguous regions that have a significant Garo population (Dutta, 2015). While armed conflict in Garo Hills is "ethnic" in that it invokes primordial claims of place, kinship, and ancestry, it is also political in that central contestations pertain to questions of political autonomy and socioeconomic development within a territorially circumscribed nation state (Tambiah, 1996).

The concept of coloniality is an important one in interrogating ethnic conflict in Northeast India. Coloniality refers to systems of power fortified during colonial rule, but which continue to shape culture, labor, and intersubjectivity, well beyond the strict limits of colonial administration. While the end of British rule witnessed an overhaul of policies and structures of governance across the vast majority of the Indian state, the Northeast region saw a continuation of British administrative policies (Baruah, 2005). This included the aforementioned colonial categories that established different layers of citizenship and belonging, which are incongruent with geopolitical realities as well as people's lived experiences. Furthermore, many communities in Northeast India also allege that the region was annexed to the Indian state without explicit consent from the people who lived there. The coloniality of the Indian state is also evident in its militarized response to ethnic conflict in Northeast India, namely through the Armed Forces (Special Power) Act of 1958 (AFSPA),[4] which grants extraordinary power and impunity to the armed forces. Derived from a draconian ordinance employed by the colonial rulers to suppress popular resistance during the Quit India Movement of 1942, the act continues to be used against those who are considered non-compliant national subjects. For example, the Assam Rifles, a major paramilitary force in the region, originated during the British administration as protection against "wild and unruly tribes" and subsequently functioned as the "the strong arm of the civil administration in extending their authority into these remote inaccessible areas" (see: http://assamrifles.gov.in). AFSPA is implicated in suspension of civil liberties and widespread violence, especially against tribal communities in Northeast India (Haksar & Hongray, 2011).

---

[3] In an effort to avoid reifying categories such as tribal and non-tribal, I focus on the material and psychosocial lived experiences associated with these labels.

[4] http://www.satp.org/satporgtp/countries/india/document/actandordinances/armed_forces_special_power_act_1958.htm

Examining Discourses of Violence from Spaces of Rupture

Growing up in the Garo Hills region of Northeast India, I continually wrestled with the (post)colonizing gaze that reduced my home community to trite characterizations such as "remote," "underdeveloped," and "disturbed." I became increasingly cognizant of the colossal gap between dominant discourses (popular, scholarly, and public policy) and the experiences and aspirations of those of us who lived those labels on a daily basis. I was drawn to the social sciences as a means of reframing ethnic violence in Northeast India by foregrounding the lived experiences and analyses of ordinary people. This foray into critical social sciences in Western academe was also marked by ruptures as I discovered the peripheral position ascribed to majority world scholars who study their own communities—the "halfie researcher" (Abu-Lughod, 1991) or "Native scholar" (Russel y Rodríguez, 1998). We are not only faced with the challenge of representing (our) communities in ethical and discursively complex ways, our invested status is also viewed as an impediment to "rigor." Thus, gaps or ruptures (rather than continuity) constitute the normative context of my scholarship. As such, I was drawn to women of Color feminist scholars who privilege and critically engage interstices and ruptures in an effort to disrupt colonialist ideology and open up possibilities for agency and transformation (e.g., Pérez, 1999; Sandoval, 1991; Villenas, 2006). For the purpose of this chapter, I will examine two sets of silences or gaps. The first pertains to silences that characterize dominant psychological conceptualizations of violence while the second examines the deep disjuncture between dominant discourses of conflict and the lived experiences of diverse constituencies in Garo Hills.

Studying the "Psychology of Violence" in (Post)colonial Contexts: Shifting Epistemic Parameters

Epistemic colonization is intertwined with ontological colonization to establish different levels of humanity (Bulhan, 1985; Connell, 2014; Santos, 2015). As such, a skeptical epistemic stance that disrupts the intellectual hegemony of the Euro-American knowledge enterprise, is an important decolonial imperative (Seedat & Suffla, 2017). A skeptical epistemic attitude requires us to interrogate mainstream knowledge claims via questions such as: Who has the authority to name and legitimize what constitutes violence? Whose concerns are excluded from dominant definitions? What are the linkages between global, state, and interpersonal violence? How do particular theories and methods naturalize forms of colonial and racial violence, such that they become difficult to name as violence? These questions not only help scrutinize the material and ideological underpinnings of how certain types of violence are legitimized, but also open up possibilities for naming and centering lived realities that are absent from dominant discourses of violence. Uncovering silenced voices and honoring multiple experiences are integral to the project of decolonizing studies of violence.

In delving into the literature on violence, establishment psychology's preoccupation with definitional consensus becomes apparent fairly quickly. A stance of epistemic skepticism behooves the question—what is at stake in disciplining definitions of violence? The aspiration for standardized nomenclature is considered integral to the "scientific" project of identifying, predicting, and controlling various antecedents and consequences of violence (e.g., Best, 2016; Hamby, 2017). This research trend not only disregards the role of coloniality in producing and sustaining violence, they also fail to account for psychology's willing complicity in perpetuating oppression and social control (e.g., racial profiling, surveillance, torture, etc.) (Bulhan, 1985; Dutta, Sonn, & Lykes, 2016). For the most part, this literature fails to examine structural or cultural violence, which obscure sociopolitical roots of inequality and oppression. This critique also applies to dominant social psychological theories of intergroup relations and intergroup conflict, that by and large, fail to foreground the violence of slavery, colonialism, imperialism, neoliberalism, xenophobia, and anti-immigrant violence in analyzing relations between "minority" and "majority" groups (e.g., Alexandre, Rosa, & Waldzus, 2016; Kaufman, 2006; Schmid, Hewstone, Küpper, Zick, & Tausch, 2014; Smith, 1999). As a consequence, some lives come to be constructed as inherently violent in order to deny the violence against them (Fine & Ruglis, 2009; Scheper-Hughes, 2004).

The terms of inclusion/exclusion associated with conceptualizations of violence have important implications for examining ethnic conflict in Northeast India. Mainstream (Euro-American) psychology has perpetuated universalizing discourses of violence within and against majority worlds, naturalizing such violence, rendering it invisible or hypervisible, and/or locating it outside the mantle of human civilization (An-Na'im, 2013; Bulhan, 1985; Das, 2002). Thus, denaturalizing violence—attending to violence as both actuality and potentiality, considering the manifold ways in which it is materialized and legitimized, as it resides in everyday life while constituting an exception from it—is crucial to the project of decolonizing studies of violence and striving for greater epistemic justice. In the context of Garo Hills, this involves studying the subtleties and complexities of power/difference through the lenses of structural and cultural violence. The efficacy of concepts of structural and cultural violence thus lie in their ability to elucidate the social machinery of oppression as well as the mechanisms that justify it (Galtung, 1990).

Over the last 14 years of my critical ethnographic research, I have used varied methods (e.g., interviews, participant observation, archival research, autoethnography, focus groups, oral histories, and participatory action research), working with a range of social actors (e.g., youth, law enforcement, armed insurgent groups, civil society organizations, and women activists) in Garo Hills to interrogate the ways in which violence is reworked through the routines of daily life and enacted through social relations and institutions (see: Dutta, 2013, 2015, 2017a). Interrogating and bringing complex, at times contradictory lived realities, into a critical analysis, I explore social reproduction of conflict as well as sites of resistance. Importantly, this work centers the voices of youth and women in Garo Hills, which tend to be excluded from dominant public discourse. Epistemic justice also requires an intentional centering of relational epistemological positions, which open up possibilities for alternative, nonhierarchical relationships while recognizing the materiality of power asymmetries (Pillow, 2003; Visweswaran, 1994). My role as a scholar activist captures a fraction—albeit a significant one—of my broader relationship to communities and contexts in Garo Hills. Given this history, many textured roles/relations have been forged with diverse constituencies, across time, and at complex intersections of gendered, racialized, and class-based experiences (see: Dutta, 2015, 2017a). Thus, this work then has been geared towards producing "messy" texts informed by a reflexivity of discomfort that "cannot be a simple story of subjects, subjectivity, and transcendence or self-indulgent tellings," but as practices of confounding disruptions—at times even a failure of our language and practices" (Pillow, 2003, p. 192).

Reframing Ethnic Conflict through the Lens of Structural and Cultural Violence

My ethnographic research on ethnic conflict in Garo Hills revealed deep and varied disjuncture between dominant discourses of conflict and the lived experiences of diverse constituencies in Garo Hills. It underscored the gap between state-sponsored ethnic identity categories (tribal and non-tribal) and people's lived experiences, especially forms of structural and cultural violence that do not squarely map on to ethnic faultlines (Dutta, 2013, 2015). Here, I draw from my critical ethnography to elucidate some of the lived experiences at these "rupturing spaces" (Pérez, 1999). This is a necessary move to avoid reinscribing the violence inherent in colonial categorizations; as Pérez (2003, p. 123) put it: "When conceptualized in certain ways, the naming of things already leaves something out, leaves something unsaid, leaves silences and gaps that must be uncovered."

"They think this place is called Garo Hills, so it is for the Garos only:" Non-tribal youth experiences Across multiple interviews and focus groups, non-tribal youth described multilayered experiences of being the ethnic other in Garo Hills, despite being born and raised there (Dutta, 2015). At times, these were manifested as exclusions from civic and community life, whether it was through tacit assumptions about participation or explicit exclusion via codification of ethnic differences. As Kavi, a non-tribal youth put it: "We can never be at par with them [Garos]. No matter what they say, in their minds we are always separate." Ethnic othering was also expressed through acts of bodily harm such as being shoved, hit, or beaten up:

> It happens everywhere, in offices, in banks, at the post office. But no one says anything. Everybody feels scorched but there is nothing to be done. No one to complain to.

Kavi's evocative statement underscores two fundamental features of being othered in Garo Hills. First it highlights its pervasiveness and ubiquity. Non-tribal youth iterated the ubiquitous risk that apparently innocuous interactions could easily precipitate into and/or aggravate physical altercations. Garo youth too agreed with this characterization. As an example, Patrick, a Garo youth acknowledged: "Honestly if I tell the truth then, yes, most of the Garo youth do not like non-tribals…if some problem or difficulty crops up then the thought that 'he is a non-tribal and I am a tribal' comes in between." Kavi's comment also points to the absence of avenues to seek redressal in the face of ethnic othering and violence. The wider society in Garo Hills is largely silent on these daily expressions of hostility and violence along ethnic fault lines (Dutta, 2015). This silence combined with institutionalized indifference serves to normalize the everyday violence experienced by non-tribal youth. Thus, it is not only actual physical violence, but entrenched mechanisms of legitimation and helplessness in the face of it, that creates a repressive environment for non-tribal youth as they go about their daily lives.

"We are not original inhabitants of this place:" Garo tribal youth experiences The narratives of Garo youth were also replete with accounts of alienation and othering (Dutta, 2013, 2015). They were the ethnic other vis-à-vis mainland India. As Patrick, a Garo youth explained: "We are not original inhabitants of this place [Indian state]. That differentiates us from mainland India. We originated in Tibet or somewhere." Most tribal groups in Northeast India trace their lineage to Mon-Khmer and Tibeto-Burman subfamilies of the larger Indo-Chinese linguistic group, and as a consequence tend to be positioned as the racialized other vis-à-vis mainland India (Bora, 2010). Additionally, many Garo youth tend to situate themselves as inferior to non-tribal communities/mainland India–often citing this perceived inferiority as the reason for widespread resentment against ethnic others (non-tribals), as exemplified in the following quotes excerpted from Garo youth narratives:

> Compared to other communities, we Garos are the most uncivilized. Even though most of us are educated now, we still have the tribal mentality of our forefathers.
>
> Non-tribals are better than us in everything! They [Garos] feel that one day they will be overrun and ruled over by non-tribals and it is this thing that makes them afraid.

My ethnographic research revealed important parallels between Garo youth accounts and official discourses of ethnic conflict. A local government official, for example, characterized ethnic conflict in Garo Hills as an "integral part of any tribal society"—arguing that "the more the society will progress, the more these demarcations will be diluted." In addition to naturalizing violence against non-tribals, such official discourses are silent on histories of marginalization of tribal communities, which serve to justify and maintain state-sanctioned violence against them. The colonial mindset of Garo youth led me to an excavation of colonial and post-colonial historiography. My research suggested that not unlike many other postcolonial contexts, the accounts of colonial administrators and missionaries became the canons of Garo history. One such publication is entitled 'The Garo Jungle Book,' written by Revd. William Carey. With chapter titles such as "The Jungle Lair," "The Wild Man at Home," and "Man-cubs at School," Carey (1919) characterized Garos as violent, hedonistic, and lazy—characterizations that got codified in postcolonial policies. The Ministry of Tribal Affairs (Government of India), for example, defines tribes in such terms as "primitive," "geographical isolation," "shyness of contact," "and "internecine feuds" (see: http://tribal.nic.in)—going on to show how violence goes beyond physicality to include assaults on self-respect and personhood (Galtung, 1990). Colonial discourses have also been reproduced in the vast majority of post-colonial scholarship *about* Northeastern tribes, tacitly representing them as racialized others who are unable to assimilate to the hegemonic idea of "India" and therefore engaged in constant conflict (e.g., Roy & Rizvi, 1990; for further critique, see: Baruah, 2005; Kikon, 2009). These discursive practices are in turn implicated in widespread stereotypes and discrimination against Northeastern tribes and are also used to justify state-sanctioned violence (Dutta, 2015).

"Everyone tries to hush it...and that is violence:" Naming gendered violence Whereas state-sanctioned violence sometimes manages to find public articulation via human rights discourses, configurations of gendered violence that do not fit into or advance dominant narratives of ethnonationalism, end up being silenced (Bora, 2010; Dutta, 2017a). My ethnographic work revealed the disjuncture between Garo women's lived experiences and the various discourses that come to bear on their lives—public discourse in Garo Hills, mainland Indian scholarly and public policy discourses, and dominant international (Western) development discourses. As Namchi, a Garo woman, pointed out, "Everyone tries to hush it...and that is violence. We may not see what is hidden but it's there. And women are suffering from it every day." Garo society is a matrilineal society in which descent is traced along the female line (Marak, 2007). The cultural importance ascribed to Garo women's reproductive roles and their custodial rights over property defy standard conceptualizations of female subordination in India (Bora, 2010; Nongbri, 2000). Thus, dominant discourses fail to consider how women's strong identification with their community (as property, as signifier, as lineage) makes them susceptible to violence, especially during periods of intensified ethnic conflict (Bora, 2010; Das, 2008; Kikon, 2005). For example, in 2014, during a period of heightened conflict, a 35-year old woman was brutally killed when she tried to resist sexual assault.[5] The Garo National Liberation Army, a major ethnonationalist group at the time, claimed responsibility, alleging that she was a police informant.[6] The collective silence following this incident was in stark contrast to widespread outrage and civil mobilization in response to recent (male) custodial deaths, which were articulated within a human rights discourse (Dutta, 2017a).

Rupturing the silences inherent in public discourse, Garo women's analysis point to the fundamental heterogeneity of violence against women—configured across socioeconomic insecurities, domestic violence, human trafficking, and

[5] Chokpot horror. (2014, June 5). *The Shillong Times*. Tura. Retrieved from http://www.theshillongtimes.com/2014/06/05/chokpot-horror/.

[6] During my interviews with members of the State Women's Commission, I learned that such counter allegations are commonly used by armed groups to justify killings.

militarism—while also tracing the cultural, institutional, and structural processes that undergird this violence (Dutta, 2017a). Such critical analyses continue to be absent from public policy discourses that frame violence in terms of discrete categories. For instance, in a recent speech[7] announcing the state government's plans in the wake of declining insurgency in Garo Hills, the Home Minister proclaimed that law enforcement would *now* "fight crimes against women, children and human trafficking." Attending to women's analyses from within rupturing spaces emphasize how violence against women located at the margins of the nation state, is configured at intersections of structures, institutions, and histories that make their experiences irreducible to any particular kind of violence. As Namchi put it,

> It's very difficult for women to live in that kind of a condition, they don't feel safe in their own community, in their own space, in their own home … So when militancy happens in villages, women have no control over the lives that they live.

"There is no one to listen, no one to help:" Poverty and dispossession in rural Garo Hills The hegemony of ethnocentric categories and associated ethnic identity politics also tend to obscure other forms of structural violence that do not squarely map on to ethnic divisions. One such issue is that of abysmal poverty and dispossession in rural Garo Hills, the most visceral accounts of which emerged in the context of interviews with young men incarcerated for their alleged affiliation to the armed insurgent group, Achik National Volunteer Council (ANVC). These young men, aged 19–24 hailed from villages in Garo Hills bordering Bangladesh, most of them having served as "foot soldiers" in the ANVC. As 23-year old Gretinson explained:

> People in my village mostly do *hajira* [daily wage earners]. This is the only means of earning our livelihood for most of us… People are quite destitute. Being daily laborers, people are also exploited… We do not receive help from any other agencies. There are no social workers. There is no one to listen, no one to help.

This was reinforced by Hevelson who drew attention to the exclusion of voices and concerns of rural communities: "Sometimes young people go to the government but their voices are suppressed… villagers are underestimated. People think little of them. For the people living in the border areas there is so much of misery." Although discourses of poverty and underdevelopment are integral to the claims of separatist groups such as ANVC (Dutta, 2013), urban communities are largely inured to the sufferings of the rural poor (Blah, 2010). While there were discursive references to "outsiders" or ethnic others laying claim to already scant resources, destitution and suffering were cited as the primary reason for joining ANVC; as Gamseng explained: "We have to fight… I have seen how people in the villages suffer… I thought I would work for my own tribe—the Garo community." Structural violence not only animated young people's motivations to join armed groups, it was also defined the state's response to armed struggles—through the now repealed

[7] Source:http://www.theshillongtimes.com/2018/03/12/gnla-surrender-marks-beginning-of-end-to-militancy-meghalaya-home-minister/.

Prevention of Terrorism Act (POTA) that allowed for prolonged detention without requiring any evidentiary threshold (Krishnan, 2004). Consequently, the youth I interviewed had been languishing in the prison as "under-trials," without any trial or legal representation in sight. For many of them, this meant an inevitable breakdown of family and family relations—the acute impoverishment constituting an insurmountable barrier to visitations.

Centering Gaps/Ruptures as Epistemic Decolonization

The gaps between official categorizations/discourses and the lived reality of youth and women in Garo Hills offer a glimpse into the ways in which everyday life is shaped by historical processes as well as by local cultural discourses. For Garo youth, the presence of a territorially circumscribed home (Garo Hills) within the nation state does not automatically guarantee a connection to the nation state. As racialized others, they are targets of state violence and face cultural assaults on their dignity and self-worth. Non-tribal youth, who live in Garo Hills and consider it home, have come to accept the multiple ways in which violence and marginality shape their daily lives. This is an example of the devastation wrought by (post)colonial practices of codifying groups and cultural practices that were not only capable of accommodating diverse local circumstances, but in fact depended on categorical opaqueness for peaceful coexistence (Kabeer, 2005; Mamdani, 1996). A legacy of historical colonial construction, state-sponsored ethnic identity categories have created shifting tiers of majority and minority groups (e.g., some national minority groups such as Garos are majority at regional levels) with differential access to resources (Abah & Okwori, 2002; Barbora, 2008). The predicament of youth in Garo Hills highlights the tensions between universal aspirations of citizenship and the particularlist claims of identity and belonging (Isin & Wood, 1999; Yuval-Davis, 2006).

Plotting the discontinuities also enable us to name forms of structural violence (poverty, dispossession, and social exclusion) that animate armed struggles in Garo Hills. Lenses of structural and cultural violence not only complicate binaries (e.g., victim/perpetrator), they also shed light on forms of inequality and oppression that have become endemic to the social fabric and are no longer questioned—such as the manifold against women. Attending to Garo women's standpoints highlights the centrality of gender in the understanding of violence and the ways in which it reveals the deep connections between the spectacular and the everyday (Das, 2008; Dutta, 2017a). In other words, they render visible the gendered social, political, and cultural dimensions that give violence its force and meaning. Taken together, the rupturing or interstitial spaces examined above, underscore the fraught character of *everyday* in contexts where violence is not an aberration of the social context but is configured by it. While the focus on "ethnic" violence imputes a certain intractability to the conditions in Garo Hills, attending to lived citizenship—the ways in which people's sociocultural and material circumstances animate their lives as citizens—from within gaps/silences offers important insights for psychosocial transformation.

Everyday Peace Praxis in Garo Hills

Violence that is embedded in the recesses of everyday life calls for a corresponding notion of peace, one that that attends to the ways in which injustices wrought by colonialism, racism, patriarchy, imperialism, and neoliberalism are woven into the fabric of everyday life (Dutta, Andzenge, & Walkling, 2016). Discursively constructed as exceptional or isolated, dominant scholarly and public policy responses to ethnic conflict in Garo Hills, fail to examine the manifold experiences of marginality and violence. Earlier, we see how the testimonies of youth and women in Garo Hills are denied access to participation in the generation of social meanings, that is, they lack epistemic authority (Fricker, 2013). As such, decolonizing notions of "peace" involves reclaiming the construct from local, national, and international gatekeepers. Not unlike the "civilizing" mission of Christian missionaries, Western liberal notions of conflict resolution, peacebuilding, and human rights are predicated on the imposition of particular configurations of state, economy, and culture that reproduce colonial violence (An-Na'im, 2013; Collins, Falcón, Lodhia, & Talcott, 2010; Haksar & Hongray, 2011; Maldonado-Torres, 2017; Turner & Kühn, 2015; Walker, 2004). Simultaneously, we have to recuperate peace from the shackles of official "peace treaties." During almost three decades of armed conflict in Garo Hills, various armed insurgent groups have entered into tripartite ceasefire agreements with the central and state governments. Characterized by mistrust, threats, and violations on all sides, these "peace talks" are anything but peaceful (see: www.satp.org). These talks tend to involve the leaders of armed insurgent groups, while members in the lower rungs form break away groups, are arrested, compelled to surrender, or killed by police/military or fellow insurgents. In these blurred cycles of armed violence, peace in the form of ceasefire or treaties mean very little to people on the ground.

Responding to these exigencies in Garo Hills, I began to explore the idea of "everyday" peace praxis as a way of reimagining peacebuilding in Garo Hills. The term *everyday* has particular implications here. First, it is responsive to local configurations of structural and cultural violence. Second, it counters dominant, top-down conceptualizations of peace that reproduce colonial relations. Finally, *everyday* peace alludes to a deliberate democratization of the notion of peace. Rather than working with pre-existing notions of peace, it is for communities to grapple with what peace means given their historical, political, and sociocultural particularities. In the context of Garo Hills, existing chasm between constitutionally guaranteed citizenship and everyday meanings and expressions of citizenship underscores the importance of citizenship as an arena for everyday peace praxis. One possible pathway for everyday peace praxis is through exploration spaces for articulating new, more inclusive, gendered meanings of citizenship that center excluded standpoints (Isin & Wood, 1999; Kabeer, 2005; Lister, 2008; Yuval-Davis, 2001). Consistent with decolonial approaches, this is not about "including" voices through incorporation or tokenism but a radical transformation in the terms of recognition and participation (Dutta, 2016). Participatory action research (PAR), with its explicit attention to praxis—reflection and action upon the world in order to transform it—represents a powerful vehicle for democratizing peace and exploring inclusive citizenship. PAR is aligned with indigenous conceptions of justice characterized by An-Na'im (2013, p. 199) as: "[H]omegrown, emerging out of local context and adapted by the people themselves to the specific conditions of their country, through practice, and not as normative ideals imposed from above."

Key tenets of PAR, with their radical Southern roots, are aligned with a decolonial agenda: rejection of the subject/object binary; recognition of subjugated or "organic" knowledge as expertise; primacy of human dignity; collective inquiry and action based on self-determined priorities; and the imperative to work toward a more humanized conception of the world (Bulhan, 1985; Fals Borda, 1979; Freire, 1982; Maldonado-Torres, 2016; Rahman, 1990). PAR has the potential to foster critical consciousness, build solidarities, and inspire collective mobilization—processes that are integral to inclusive citizenship practices (Kabeer, 2005; Yuval-Davis, 2007). Toward that end, I worked with a group of ten youth across different tribal and non-tribal ethnicities on a PAR project in Garo Hills. As a facilitator and critical friend (Evans, 2015), I framed the project as an opportunity for youth to interrogate pressing issues in Garo Hills. The specific issues, research questions, methods, and documentation emerged out of the scaffolded PAR process and were crucially shaped by youth. We worked together to build a community of researchers where the participating youth honed a critical lens and acquired research and documentation skills to aid their efforts. As part of the PAR project, youth researchers examined archival data and interviewed stakeholders such historians, educators, their local representatives to the state legislative assembly in an effort to acquire a critical understanding of contemporary issues. They also designed and conducted a survey to examine the problems faced by ordinary citizens in Garo Hills as well as their aspirations. This idea grew out of a collective critique as they noted the absence of spaces for ordinary people to express their concerns—as John, a youth researcher put it: a space "to highlight the problems faced by people of all communities present in the area and a medium that bring them together."

The youth researchers conducted surveys in various urban localities, often using the survey to initiate critical conversations with residents. We analyzed the data using a critical dialogic process that is associated with the promotion of critical consciousness (Montero, 2009). Calling themselves *Voices,* the youth PAR team disseminated their findings at local venues, such as relevant community and university events and the local television channel. Since detailing the entire PAR project (see: Dutta, 2017b) is beyond the scope of this chapter, I will limit my discussion to the impact of the citizens opinion survey—impact that accrued from the process of designing and conducting the survey as well as dissemination of findings at a large public event in Garo Hills. Specifically, I will highlight how young people subverted local cultures of ethnic othering/exclusion while successfully countering tokenistic modes of *participation* that characterize the approach of dominant national actors.

Radical Participation and Inclusive Citizenship Practice as Everyday Peace Praxis

The citizens' opinion survey served as the vehicle through which young people were able to *speak truth to power* (Martín-Baró, 1996). As Trish, a youth researcher explained: "We can always criticize people who are equal in rank to us. But criticizing someone or commenting about someone who is above us in position is very difficult. And we successfully achieved that." The survey conferred a critical collective power that allowed young people to engage in social critique without being overly concerned about retaliation. Remarkably, youth researchers not only asserted their own voices, they also positioned themselves as *representatives of ordinary people's voices*. In Dia's words:

> We were nervous but one thing that strikes me is that I was feeling really good. The local people's opinions—they have conveyed their message to us and we are conveying it to the governor and to other people sitting there. And at that moment I felt like we can really bring change to the society.

The impact of the youth panel was not lost on the wider community, as exemplified in comments such as "I never thought our youth is capable of this kind of constructive work" and "Your initiative is a lesson to older folks."

It is important to note that "participation" is not a given in PAR. For PAR to serve as a vehicle for everyday peacebuilding, we have to scrutinize the politics of participation in specific contexts. In the context of Garo Hills, youth concerns tend to be dismissed, with youth political participation viewed as inappropriate and disruptive (Dutta, 2013). The local context is also mired in divisive ethnic identity politics that exclude non-tribal youth from civic participation. These existing challenges were compounded by the theme of the public event that constituted a major venue for dissemination of findings from the PAR project. Entitled, "The Green Hills of our Forefathers: Revisiting the Vision," the event was organized by a Delhi-based non-governmental organization (NGO). The NGO identified four "visionary leaders" of Garo Hills as well as local scholars who would speak to their relevance of their visions for "modernization" and "reform." Additionally, the NGO invited "eminent constitutional experts, scholars in the concerned fields and social activists from the rest of the country" to attend the event as "observers."[8] Their purported role was to "appreciate the problems that torment the people from within" in order to make policy recommendations. Beneath the ruse of seeking local participation, is a

[8] The Green Hills of Our Forefathers: Revisiting the Vision (Feb 15, 2010). Concept paper on the proposed workshop at Tura covering the Garo Hills Region of Meghalaya. Divya Jeevan Foundation, New Delhi, India.

paternalistic and tokenistic approach that relegates Northeastern tribal communities as objects of "reform," distinguishing them from "experts" from "the rest of the country." The predetermined theme had the effect of discouraging perspectives that did not fit into the expressed vision. The youth researchers, for example, were concerned about the risk of critiquing the perspectives of identified visionaries, one of whom was a powerful politician. Furthermore, the focus on "forefathers" served to deepen existing ethnic fault lines that threatened to delegitimize participation of non-tribal youth. For instance, Sujit, a non-tribal youth expressed: "We felt like they'll [Garo youth] be wondering what we're doing here… it was about *forefathers*, so maybe they wondered what *we* would know about *their* forefathers." As a group, we devoted considerable space to discussing and working through these concerns. In the spirit of PAR, the youth group decided to put together a panel at the event that focused on the present, and on the concerns of ordinary people in Garo Hills—thus subverting the NGO's stated agenda.

As the youth engaged in collective research tasks, they began to view their group as a resource and a source of support. Tina captured the dominant sentiment in the group:

> What we do may not be perfect but whatever we do, it is as a group. When it became clear that we were representing our group and not on our own, it felt good. And if I make a mistake, my group members will tell me and I will get a chance to correct myself. It is a learning process and I thought, yeah, this is a safe ground to learn.

This shift was particularly pronounced for non-tribal youth, given their precarious position as ethnic others in Garo Hills. Bijoy, a non-tribal youth, shared how he eventually changed his initial decision to "remain in the background" out of safety concerns, when he realized that "we all care about our place, that is, Garo Hills, and the tribal versus non-tribal distinction became less important." Forging connections between personal and collective experiences of "ordinary citizens" (rather than specific ethnic groups), this engagement afforded young people an opportunity to disrupt long-standing patterns of ethnic antagonisms. The youth researchers essentially debunked the "collective lie" or the gap between ordinary people's lived experiences and "official" discourses (Martín-Baró, 1996). In the process, they **co-constructed a fledgling, yet significant collective identity**—one that is premised on **solidarity** around a critical recognition of shared concerns; solidarity as "the capacity to identify with others and to act in unity with them in their claims for justice and recognition" (Kabeer, 2005, p. 7). The youth group began to articulate a **politics of possibility** that integrated both personal and collective agency—exemplified in the following statement from Pansy:

> After being a part of this project, discussing and working together for our community, it has become a part of our lives, something to look forward to…the small project was so interesting and successful—imagine what we can do as a group!

The PAR project, specifically the citizens' opinion survey—from its genesis as an idea to its implementation and beyond—established integrative, crosscutting ties that bypassed ethnic fault lines; in doing so, it began to address critical psychosocial and political dimensions of belonging while fundamentally challenging the notion of intractability ascribed to protracted ethnic conflict. Rendering this sort of clarity or intelligibility is an important instrument for decolonization in the face of systematic exclusion, concealment, and distortion though which coloniality tends to operate (Maldonado-Torres, 2016).

Youth researchers' testimonies also indicated a growing **sense of agency** or the belief that one can act as well as actually acting as citizens (e.g., engaging in social critique, making recommendations to policy makers), which reinforced the politics of possibility (Kabeer, 2005; Yuval-Davis, 2007). This is noteworthy given the fundamental connection between collective agency and inclusive citizenship practices (Isin & Wood, 1999; Lister, 2008). This intentional, collective move on the part of young people to contest their marginalized and subordinated status marks their shift from subject to citizen. Thus, the group processes may be understood through the lens of "horizontal" citizenship that prioritizes relationships *between* citizens in contrast to the more traditional top-down relationship between state and individual (Kabeer, 2005, p. 23). This view of citizenship is especially relevant in contexts such as Garo Hills where the state is, by and large, unresponsive to varied needs and concerns of ordinary people. Furthermore, a horizontal view of citizenship focuses on collective actions of those citizens who are disenfranchised by the state. As such, it has important implications for politics of recognition or respect; that is, how people perceive themselves as citizens and how/whether they are recognized by others shape how they act to claim their citizenship rights (Abah & Okwori, 2002; Lister, 2008). In thus subverting established patterns of marginality and social exclusion while demonstrating more inclusive collective identifications, *Voices* articulated a concrete possibility of everyday peace praxis in Garo Hills.

Conclusion

In this chapter, I argue for the need to move away from normative notions of violence that tend to spectacularize, demanding instead a focus on the "everyday". I illustrate this through an epistemic reconstruction of ethnic conflict in Garo Hills by privileging ruptures between dominant discourses (scholarly, popular, and public policy) and lived experiences and analyses of local people. I have shown how attending to rupturing spaces is integral to a decolonial analysis of violence: it helps us denaturalize "ethnic" violence in postcolonial contexts via considerations of structural and cultural violence (e.g., cultural and historiographical marginalization, dispossession, and social exclusion) while centering silenced or subjugated perspectives. We see how young people's substantive and affective expressions of nation and community demonstrate their embeddedness in everyday entanglements of authority and alterity. This analysis not only offers a nuanced and discursively complex understanding of protracted ethnic conflict in Garo Hills, it also provides critical insights into everyday peacebuilding possibilities predicated on more democratic and just alternatives to Western liberal notions of peace. Drawing upon a youth PAR project, the chapter illustrates how radical participation can disrupt dominant patterns of exclusion, allowing youth to forge a collective consciousness and mitigate the damage to civil society wrought by everyday violence. Unsettling universalized notions of peace, everyday peace praxis emphasizes the value of local constructions of peace that embody the social positions, lived experiences, and geopolitical and affective realities of local communities. This chapter is certainly not without its limitations. While epistemic justice has been a critical focus of the paper, it is crucial to note that it is a necessary but not sufficient condition for decoloniality. In writing this chapter, I have also been uncomfortably cognizant of those details that were excluded for the sake of narrative coherence. There is an inherent tension in trying to articulate ruptures through a neatly woven narrative, trying to make the different parts cohere. As important as it is to contend with the substantive aspects of decolonial analysis, it is also crucial to continually grapple with the politics and ethics of writing and representation—a reminder that "being on the edge" of ease/unease is an existential condition of engaging in decolonial work.

References

Abah, O. S., & Okwori, J. Z. (2002). Agendas in encountering citizens in the Nigerian context. *IDS Bulletin, 33*(2), 24–30.

Abu-Lughod, L. (1991). Writing against culture. In R. Fox (Ed.), *Recapturing anthropology: Working in the present*. Santa Fe, NM: School of American Research Press.

Alexandre, J., Rosa, M., & Waldzus, S. (2016). Intergroup relations and strategies of minorities. In *The social developmental construction of violence and intergroup conflict* (pp. 55–82). New York, NY: Springer.

An-Na'im, A. A. (2013). Editorial note: From the neocolonial 'Transitional' to indigenous formations of justice. *International Journal of Transitional Justice, 7*(2), 197–204.

Bal, E. (2007). *They ask if we eat frogs: Garo ethnicity in Bangladesh*. Singapore: Institute of Southeast Asian Studies.

Barbora, S. (2008). Autonomous districts and/or ethnic homelands: An ethnographic account of the genesis of political violence in Assam (North-East India) against the normative frame of the Indian constitution. *International Journal on Minority and Group Rights, 15*, 313–334.

Baruah, S. (2005). *Durable disorder: Understanding the politics of Northeast India*. New York: Oxford University Press.

Best, J. (2016). The social construction of violence. In C. A. Cuevas & C. M. Rennison (Eds.), *The Wiley handbook on the psychology of violence* (pp. 86–99). New York, NY: Wiley.

Blah, T. (2010, August 3). *Meghalaya silhouettes—The socio-cultural profile*. The Shillong Times: Tura Edition.

Bora, P. (2010). Between the human, the citizen and the tribal: Reading feminist politics in India's Northeast. *International Feminist Journal of Politics, 12*, 341–360.

Bulhan, H. A. (1985). *Frantz Fanon and the psychology of oppression*. New York, NY: Plenum Press.

Carey, W. (1919). *A Garo Jungle Book or the mission to the Garos of Assam*. Philadelphia, PA: The Judson Press.

Collins, D., Falcón, S., Lodhia, S., & Talcott, M. (2010). New directions in feminism and human rights: An introduction. *International Feminist Journal of Politics, 12*, 298–318.

Connell, R. (2014). Using southern theory: Decolonizing social thought in theory, research and application. *Planning Theory, 13*(2), 210–223.

Das, V. (2002). Violence and translation. *Anthropological Quarterly, 75*(1), 105–112.

Das, V. (2008). Violence, gender, and subjectivity. *Annual Review of Anthropology, 37*, 283–299.

Dutta, U. (2013). Beyond the ethnicity question: Constructing and enacting more inclusive identities. In J. Prodhani & R. S. Thakur (Eds.), *Culture, ethnicity, and identity: A reader* (pp. 290–299). Guwahati, India: DVS Publications.

Dutta, U. (2015). The long way home: The vicissitudes of belonging and otherness in Northeast India. *Qualitative Inquiry, 21*(2), 161–172.

Dutta, U. (2016). Prioritizing the local in an era of globalization: A proposal for decentering community psychology. *American Journal of Community Psychology, 58*(3–4), 329–338.

Dutta, U. (2017a). The everyday and the exceptional: Rethinking gendered violence and human rights in Garo Hills, India. In S. Grabe (Ed.), *Women's human rights: A social psychological perspective on resistance, liberation, and justice* (pp. 179–204). New York, NY: Oxford University Press.

Dutta, U. (2017b). Creating inclusive identity narratives through participatory action research. *Journal of Community & Applied Social Psychology, 27*(6), 476–488.

Dutta, U., Andzenge, A. K., & Walkling, K. (2016). The everyday peace project: An innovative approach to peace pedagogy. *Journal of Peace Education, 13*(1), 79–104.

Dutta, U., Sonn, C. C., & Lykes, M. B. (2016). Situating and contesting structural violence in community-based research and action. *Community Psychology in Global Perspective, 2*(2), 1–20.

Evans, S. D. (2015). The community psychologist as critical friend: Promoting critical community praxis. *Journal of Community & Applied Social Psychology, 25*(4), 355–368.

Fals Borda, O. (1979). Investigating reality in order to transform it: The Colombian experience. *Dialectical Anthropology, 4*(1), 33–55.

Fine, M., & Ruglis, J. (2009). Circuits and consequences of dispossession: The racialized realignment of the public sphere for US youth. *Transforming Anthropology, 17*(1), 20–33.

Freire, P. (1982). Creating alternative research methods: Learning to do it by doing it. In B. Hall, A. Gillette, & R. Tandon (Eds.), *Creating knowledge: A monopoly? Participatory research in development* (pp. 29–37). New Delhi, India: Society for Participatory Research in Asia.

Fricker, M. (2013). Epistemic justice as a condition of political freedom? *Synthese, 190*(7), 1317–1332.

Galtung, J. (1990). Cultural violence. *Journal of Peace Research, 27*(3), 291–305.

Haksar, N., & Hongray, S. M. (2011). *The judgement that never came: Army rule in north east India*. New Delhi, India: Chicken Neck.

Hamby, S. (2017). On defining violence, and why it matters. *Psychology of Violence, 7*(2), 167–180.

Isin, E., & Wood, P. (1999). *Citizenship and identity*. Thousand Oaks, CA: Sage.

Kabeer, N. (2005). Introduction: The search for inclusive citizenship: Meanings and expressions in an interconnected world. In N. Kabeer (Ed.), *Inclusive citizenship: Meanings and expressions* (pp. 1–27). London, UK: Zed.

Kaufman, S. J. (2006). Symbolic politics or rational choice? Testing theories of extreme ethnic violence. *International Security, 30*(4), 45–86.

Kikon, D. (2005). Women's rights and autonomy in Naga politics and society. In *1st National Conference on Critical Thinking, Kolkata* (pp. 29–30).

Kikon, D. (2009). The predicament of justice: Fifty years of armed forces special powers act in India. *Contemporary South Asia, 17*, 271–282.

Krishnan, J. K. (2004). India's Patriot Act: POTA and the impact on civil liberties in the world's largest democracy. *Law & Inequality, 22*(2), 265–300.

Lister, R. (2008). Inclusive citizenship, gender and poverty: Some implications for education for citizenship. *Citizenship Teaching and Learning, 4*(1), 3–19.

Maldonado-Torres, N. (2016, October 23). *Outline of ten theses on coloniality and decoloniality*. Retrieved from http://fondationfrantzfanon.com/IMG/pdf/maldonado-torres_outline_of_ten_theses-10.23.16_.pdf

Maldonado-Torres, N. (2017). On the coloniality of human rights. *Revista Crítica de Ciências Sociais, 114*, 117–136.

Mamdani, M. (1996). *Citizen and subject: Contemporary Africa and the legacy of late colonialism*. Princeton, NJ: Princeton University Press.

Marak, C. R. (2007). The role of the mahari in A'chik society: Change and continuity. *SouthAsia: Journal of South Asian Studies, 30*(3), 531–540.
Martín-Baró, I. (1996). *Writings for a liberation psychology*. Cambridge, MA: Harvard University Press.
Montero, M. (2009). Community action and research as citizenship construction. *American Journal of Community Psychology, 43*, 149–161.
Nongbri, T. (2000). Khasi women and matriliny: Transformations in gender relations. *Gender, Technology and Development, 4*(3), 359–395.
Pérez, E. (1999). *The decolonial imaginary: Writing Chicanas into history*. Bloomington, IN: Indiana University Press.
Pérez, E. (2003). Queering the borderlands: The challenges of excavating the invisible and unheard. *Frontiers: A Journal of Women Studies, 24*(2/3), 122–131.
Pillow, W. (2003). Confession, catharsis, or cure? Rethinking the uses of reflexivity as methodological power in qualitative research. *International Journal of Qualitative Studies in Education, 16*(2), 175–196.
Rahman, M. A. (1990). The case of the third world: People's self-development. *Community Development Journal, 25*(4), 307–314.
Roy, S., & Rizvi, S. H. M. (1990). *Tribal customary laws of North-East India*. South Asia Books.
Russel y Rodríguez, M. (1998). Confronting anthropology's silencing praxis: Speaking of/from a Chicana consciousness. *Qualitative Inquiry, 4*(1), 15–40.
Sandoval, C. (1991). US third world feminism: The theory and method of oppositional consciousness in the postmodern world. *Genders, 10*, 1–24.
Santos, B. (2015). *Epistemologies of the south: Justice against epistemicide*. New York: Routledge.
Scheper-Hughes, N. (2004). Dangerous and endangered youth: Social structures and determinants of violence. *Annals of the New York Academy of Sciences, 1036*(1), 13–46.
Schmid, K., Hewstone, M., Küpper, B., Zick, A., & Tausch, N. (2014). Reducing aggressive intergroup action tendencies: Effects of intergroup contact via perceived intergroup threat. *Aggressive Behavior, 40*(3), 250–262.
Seedat, M., & Suffla, S. (2017). Community psychology and its (dis) contents, archival legacies and decolonisation. *South Africa Journal of Psychology, 47*(4), 421–431.
Singha, K., & Nayak, P. (2015). *Reconsidering ethnic-based-autonomy movements in Meghalaya: An analysis*. Retrieved from http://mpra.ub.uni-muenchen.de/64745/
Smith, M. B. (1999). Political psychology and peace: A half-century perspective. *Peace and Conflict: Journal of Peace Psychology, 5*(1), 1.
Tambiah, S. (1996). *Leveling crowds*. Berkeley, CA: University of California Press.
Turner, M., & Kühn, F. P. (2015). *The politics of international intervention: The tyranny of peace*. Routledge.
Villenas, S. (2006). Latina/Chicana feminist postcolonialities: Un/tracking educational actors' interventions. *International Journal of Qualitative Studies in Education, 19*(5), 659–672.
Visweswaran, K. (1994). *Fictions of feminist ethnography*. Minneapolis, MN: University of Minnesota Press.
Walker, P. O. (2004). Decolonizing conflict resolution: Addressing the ontological violence of westernization. *The American Indian Quarterly, 28*(3–4), 527–549.
Yuval-Davis, N. (2001). Contemporary agenda for the study of ethnicity. *Ethnicities, 1*(1), 11–13.
Yuval-Davis, N. (2006). Belonging and the politics of belonging. *Patterns of Prejudice, 40*(3), 197–214.
Yuval-Davis, N. (2007). Intersectionality, citizenship and contemporary politics of belonging. *Critical Review of International Social and Political Philosophy, 10*(4), 561–574.

Decoloniality in Being Māori and Community Psychologists: Advancing an Evolving and Culturally-Situated Approach

Mohi Rua, Shiloh Groot, Darrin Hodgetts, Linda Waimarie Nikora, Bridgette Masters-Awatere, Pita King, Rolinda Karapu, and Neville Robertson

In abstract discussions of colonisation, it has become fashionable in some circles to invoke contemporary 'post-colonial' settings. An unintended inference from the use of this construct is that colonisation is somehow over and we have entered a new post-colonial epoch. As McGuinne (2013, para. 2) astutely observes for those who claim to advance 'critical' approaches to post-colonial theory and research, "…decolonisation is an interesting debate to be had over a glass of white w(h)ine, to the rest of the world, decolonisation is life". As subjects of continued colonial practices, we find notions of post-coloniality hard to stomach. In order to realise decolonisation we must first understand that colonisation has not ended. Today, colonisation is reproduced in everyday instances when Māori[1] women have to adopt a subservient persona that is 'acceptable' to colonial institutions in order to face a little less stigma, denigration and racism when trying to access support to feed their children (King, Rua, & Hodgetts, 2017). Such women would not have to beg for help if Māori still held guardianship over their traditional resources (land, forests

[1] Indigenous peoples of Aotearoa/New Zealand.

M. Rua (✉) · B. Masters-Awatere · N. Robertson
Māori & Psychology Research Unit, University of Waikato, Hamilton, New Zealand
e-mail: mohi.rua@waikato.ac.nz

S. Groot
School of Psychology, University of Auckland, Auckland, New Zealand

D. Hodgetts · P. King
School of Psychology, Massey University, Wellington, New Zealand

L. W. Nikora
Indigenous Studies, University of Auckland, Auckland, New Zealand

R. Karapu
Te Whakaruruhau-Waikato Women's Refuge, Hamilton, New Zealand

G. Stevens, C. C. Sonn (eds.), *Decoloniality and Epistemic Justice in Contemporary Community Psychology*, Community Psychology,
https://doi.org/10.1007/978-3-030-72220-3_10

and sea), which were confiscated illegally by the settler society to service their own needs.

Māori live with constant reminders of the ongoing impacts of colonisation in terms of the disproportionate rates of suicide, domestic violence, homicide, substance abuse and addiction, incarceration, hospitalisation, children taken into state care, mental illness, homelessness, and ill health in comparison to Aotearoa's settler society (Ministry of Health, 2015). A consequence of colonisation for many is displacement from everything that makes us who we are as Māori. Colonisation establishes a landscape of dominance within which our lives are shaped and scarred and within which struggles for survival and wellness is endured and borne across generations.

In response, processes of decolonisation occur in Māori communities on a daily basis through a range of socio-cultural practices that open up spaces for us to reproduce and reaffirm our culture and ways of being (Hutchings & Lee-Morgan, 2016; King, Hodgetts, Rua, & Morgan, 2017; Rua, Hodgetts, & Stolte, 2017). These efforts are often lost on psychologists (including many from critical and community traditions) who are more fixated on documenting the many negative issues impacting upon us. Attempts to engage with our strengths, knowledge frameworks, histories, capacities, and solutions are much less prevalent and often progressed by us, as indigenous peoples, for our own needs (Groot, Rua, Awatere-Masters, Dudgeon, & Garvey, 2012; Waitoki & Levy, 2016). Further, some journals such as Culture and Psychology, often invisiblize and damage our collective efforts when they provide platforms for ill-informed commentators to question the very existence of indigenous psychologies (Jahoda, 2016). To deny us our own psychologies is to deny our existence, ways of understanding people, and contributions to the plurality of the human condition. Such challenges to indigenous psychologies from the global north constitute continuations of colonial practices of knowledge suppression, epistemic violence and denigration towards colonised peoples.

Astute readers may detect a sense of frustration and urgency in the opening passages of this chapter. This is because our people are dying whilst disciplines like psychology fixate on measuring our problems or debating the intricacies of the latest theoretical fashion are of little use in our struggles. We do not have the luxury of endlessly pontificating about, for example, the nature of epistemology or to get our theory and methods 'right' before we even bother to engage with our communities (Watkins & Shulman, 2008). We do not see how hair-splitting over whether we rebranded community psychology as *critical community psychology* or *Indigenous community psychology* helps address the legacies of colonisation that populate our communities right now. As scholarly activists, it is difficult to dwell in a discipline that prioritises abstract insular debates, a fixation on the intricacies of measuring things that do not exist, and the global expansion of a psychology of irrelevance. The glacial indifference and ineptitude of WEIRD (Western Eurocentric Industrialised Rich and Democratic) psychology (Henrich, Heine, & Norenzayan, 2010) in meeting the needs of our communities must be challenged and displaced.

In writing these passages, we are not trying to be anti-theory, anti-intellectual or inward looking. As community psychologists holding privileged positions within the academy, we feel a desperate responsibility to prioritise and address the intergenerational grief and pain endured by our people. As Māori, we cannot afford to dwell within a discipline or institutions and occupy our time with playing the disciplinary game, because our people die unnecessarily every day. Our work must disrupt the complacency which can come with the comforts of being an academic, and forge a stronger path towards realising our core aspiration for Māori *to live well, to leave improved legacies for future generations, and to remain distinctly Māori.* As we will demonstrate later in this chapter, collaborations with Māori service providers and community organisations are crucial for this agenda as they work at the coalface of Māori development initiatives. We also draw strength in the knowledge that we are not alone in seeking more relevant psychologies, which can inform our efforts to achieve this core aspiration. Bringing together conceptual insights from Indigenous psychology internationally (Enriquez, 1992; Sonn, Rua, & Quayle, 2019), liberation-orientated psychologies (Seedat, 1997; Watkins & Shulman, 2008); Critical Race Theory (Crenshaw, 1989; Hill Collins, 2000), and feminism (Hooks, 1994; Te Awekotuku, 1991) is crucial for responding effectively to processes of colonisation, hardship and dispossession in our context.

In this chapter, we hope to inform the work of like-minded scholarly activists who share our burning sense of urgency in addressing the needs of our communities and for decolonising psychology and society-at-large. Below we reflect on the need to 'reside' within culturally-patterned community relationships and to work in partnership with our communities to help us all realise better lives. As we will show, such 'residing' propels us towards challenging community psychology and pushing beyond the familiarity and mimicry of imported theories, methods and analytic frames. We will illustrate the potency of residing by reflecting on a 32-year relationship between staff and students from the School of Psychology at the University of Waikato with Te Whakaruruhau (Waikato Women's Refuge), a Māori-based domestic violence provider. Our orientation towards Te Whakaruruhau is informed by the Māori proverb 'He toka tū moana' (the rock of stability that dissipates the ocean's force). This agency is a primary source of strength for women and children who are the victims of family violence as well as many of the men who have perpetuated that violence. Before reflecting upon this collaboration with Te Whakaruruhau let us spend a little more time considering the disruption of colonial practices in our research and action.

Disrupting Coloniality through Research and Participative Community Engagement

We take the very real implications of community and scholarly efforts towards decolonisation extremely seriously. To quote Tuck and Yang (2012), "decoloni[s]ation is not a metaphor". Decolonisation signifies acts of breaking free from colonial ways of thinking and acting. Central are efforts to reveal, conceptualise and challenge oppressive colonial power structures that have benefited colonising societies and driven attempts to annihilate Indigenous peoples around the world. Decolonisation (or indigenisation) provides a pathway for recovery, the re-establishment and legitimacy of cultural frameworks and the assertion of rights—deeply rooted in Indigenous ways of knowing, being and doing. Such decolonisation manifests itself in a multitude of different ways, ranging from acts as simple as assertions of Indigenous identities, to the establishment of Indigenous immersion schools[2] and research centres, to redesigning domestic violence services in accordance with the worldviews and practices of Indigenous groups.

Research produced by Māori scholars and allied colleagues is often subversive and exemplifies scholarship that seeks to disrupt coloniality and to advance processes of decolonisation. As Smith (1999, p. 198) argues: "How can research ever address our needs as Indigenous people if our questions are never taken seriously? It was as if the community's questions were never heard, simply passed over, silenced". Writing on decolonizing methodologies from a Māori perspective, Smith (1999) has delineated the methods and purposes of research in Indigenous communities: research to assist in claiming resources; testimonies, storytelling, and remembering to claim and speak about painful events and histories; and research that celebrates survival and resilience and that revitalizes language, arts, and cultural practices. Communities beset by various forms of oppression, whose members have suffered from a diminished sense of themselves through racism and classism, use such research to not only nurture community understandings, but to help preserve cultural practices (Land, 2015; Watkins & Shulman, 2008). Like others in this book (see 'Disrupting the canon' chapter), we emphasize the importance of inserting ourselves in the discipline of psychology through our own knowledge production processes in order to disrupt the hegemonic colonial practices that remain evident in WEIRD psychology and to better serve the needs of our communities.

Breaking from the top down practices of WEIRD psychology, dialogue with community members throughout the research process maximizes the potential for knowledge generation, theory building and methodological innovation to result in action that helps the people with whom we engage. As Tandon writes (cited in Watkins & Shulman, 2008, p. 283), "Dialogue becomes the vehicle for critical consciousness and praxis … [It] acts as a method to integrate inquiry and intervention". The opening of liminal spaces for dialogue is foundational to the cultivation of an engaged, community centred and relevant research and action (Watkins & Shulman, 2008). Likewise, central to the indigenizing of psychology in Aotearoa/New Zealand are efforts to develop aspects of our discipline in which Māori people can recognize themselves and their beliefs and everyday practices (Nikora et al., 2017; Waitoki & Levy, 2016). More broadly, an exciting aspect of many Indigenous psychologies (including ours) is the development of appropriate and relevant psychological knowledge. These tasks require us to draw upon and re-purpose, where relevant,

[2] In Aotearoa we have early childhood centers (5 years and under), primary schools (5–12 years) and secondary schools (13–17 years) where the entire curriculum is taught in the Māori language (te reo Maori).

international research and teaching resources from psychologies responsive to the context of our Indigenous knowledge and evolving relationships in society (Nikora et al., 2017). Such work is important for the survival, uniqueness and heritage of our local ways of being and helping each other.

In many respects, our work is a response to the pressing need to cultivate psychologies that speak to local and diverse contexts that feature issues of violence and exploitation, which undermine human flourishing (Watkins & Shulman, 2008). Central to our evolving approach are efforts to build solidarity and dialogues for change within and across different communities and professional groups. As Walia (2013, p. 248) writes, "Decolonization is more than a struggle against power and control; it is also the reimagining and generating of alternative institutions and relations". Associated shifts in power from centre to margin, from colonising to indigenising/decolonising research, necessitate the democratization of psychological knowledge production and implementation (Land, 2015; Watkins & Shulman, 2008). This democratization is not possible unless researchers schooled in the academy step down from the hegemonic positions of detached experts. It is possible when we accompany communities (often for considerable periods-of-time) that have invited us to be part of their efforts at realising better lives. It also involves working within our own communities and enacting the ethical obligations that come with the increased accountabilities of being both researchers and community members. Central to such work are efforts to "studying up" by reversing the trend of making people less fortunate than ourselves the abject/object of knowledge production. In order to build better and more secure futures together, solidarities have to be formed out of the ashes of the violent past and present, which have also dominated the production of knowledge about us in psychology. This is difficult, often intimidating and painful work. It requires us to re-imagine how we go about our work together with people who populate both our professional and personal lives.

Residing in Community

When we refer to 'residing in community', we are calling into view the relationships we have with each other and our connections to particular groups with whom our relationships are both professional and personal. It invokes our collective humanity as people working with people on issues of mutual concern. It also raises the relational nature of Māori ways of being and the expectations of connection that we often enact with one another. Residing in community positions the psychologist as part of the community and not as an expert who resides apart from the collective. For us, such residing means to live, to make a home, to share, and to remain with other people for long periods of time.

Guiding our decolonised engagements and residence within the community are a set of core Māori cultural values and expectations. These concepts can be used to form the foundations for an enacted Indigenous psychology that is anchored in our cultural ways of being, understanding and behaving towards each other. In many respects, these Māori values and expectations constitute a culturally patterned system of mutual support and engagement. *Manaakitanga* refers to our duty to care for relationships and requires us to engage as ethical colleagues engaged in joint ventures with agency staff and community members. Enacting Manaakitanga raises the importance of *Whānaungatanga*, or engaged relationships with each other. In turn, Whānaungatanga raises the importance of *Kotahitanga*, or a group reaching unity through consensus. Our culturally informed engagements within the community can also invoke notions of *Whakaiti*, or service to others with humility. In working with others we also stress the importance of *Hūmarie*, or acting with gentleness and kindness where possible. Briefly, what such core concepts offer to the world is a psychology that can strengthen our collective humanity and which offers alternative ways of being with each other and knowing the world.

For Māori, being a person means existing through relational entanglements, which are navigated in accordance with these core cultural values. It involves embracing our shared humanity and responding to the cultural expectations and obligations that texture the places in which we reside as well as the metaphysical worlds that are at play in our lives. Everything and everyone is related and interdependent in our world. This means that we have a primary duty of care for others, particularly those with whom we are related and/or reside (Mead, 2003). All Māori can lay claim to multiple descent lines through maternal and paternal lines that echo back to cosmological origins and ripple outwards across generations in particular places. Kinship networks provide a space for selves to rest and renew, and when one part of the network is for some reason stressed and burdened, the network reconfigures as we attempt to meet our duties of care. Our state of fundamentally interconnected being within which friendship and professional networks are also interwoven within kinship networks can be difficult for people operating from other cultural psychologies to unravel and fathom (King, Hodgetts, et al., 2017; King, Rua, et al., 2017; Rua et al., 2017). James Ritchie (1992), in his early studies with rural Māori communities observed that everyone was in some way connected and when queried could relate themselves across time to other people, places, events, workplaces, scandals and celebrations. While Māori mostly live in urban centres now, away from rural tribal homelands, such relatedness remains strongly evident in our everyday lives together and collective ways of doing things.

Our relationships are a vital source of resilience, resistance and respite. Moreover, the supportive relationships from which we draw our strength extend beyond kinship into additional shared histories with those around us at work and play. As neoliberal capitalism dominates our contemporary realities with paid work and education consuming large amounts of our time and energy. Whilst necessity for survival in the contemporary world, we refuse to be reduced to the self-interested lonely thinkers that come with such economic models and the WEIRD psychologies that accompany them (King, Hodgetts, et al., 2017; King, Rua, et al., 2017). Even with the challenges of oppressive work environments and economic exclusions faced by many, as Māori we never relinquished our aspiration to support each other and to change our circumstances. In this context, residing in community and being useful to others requires a solemn commitment to a shared aspiration for a better life for our people. As members of the communities in which we reside and as scholarly activists we take this commitment seriously and are motivated to do so by the desperation of the many with whom we dwell. In sum, to live well together means to remain resistant to the power of colonisation, a power that seeks to remove us from everything that makes us who we are and to leave instead a hollow shell and ravaged community.

In the exemplar that follows we demonstrate how it is possible to reside in community and simultaneously make ongoing and meaningful contributions as scholarly activists alongside others who are experts in their own work. This exemplar is one of enduring relationships that locates the people involved within a continuous web of entwined futures. What we seek to document here is not simply a stream of work restricted to isolated research projects. It is a way of engaging that demands continuity, accountability and compassion, and which is ongoing. It is an indigenised form of participative action research that supports community self-presentation and self-determination (*cf.*, Gierbolini, this collection), and much more. It is a form of action research that transcends the dichotomy between scholars, service providers and communities and entangles us within shared research practices, service delivery, and friendships. As will be exemplified with reference to our collaborations with Te Whakaruruhau, we have created a joint history that makes meaningful action to support whānau (immediate and extended family) more doable. In this context, research becomes a secondary activity to the imperatives of being ourselves as Māori through enactments of Whānaungatanga, Manaakitanga, Kotahitanga, Whakaiti and Hūmarie.

Residing with Te Whakaruruhau and Engaging Together Beyond Community

By way of further background to the exemplar, the first Māori women's refuge (Te Whakaruruhau) was established in Hamilton in 1986. This development resulted from the efforts of a group of local Māori women working to save lives and reduce the harm caused by domestic violence. Initially, the service provided a 24-h crisis response team and safe housing. Over 32 years, this safe housing has grown from a one bedroom flat into six safe houses across the city. Te Whakaruruhau rapidly developed a reputation for championing innovative and research based approaches to stopping violence that are embedded within a Māori worldview and associated cultural values. This approach includes breaking with the dictates of white feminism and to involve the whole whānau (immediate and extended family, including where possible men who have been violent) as a Māori cultural way of creating sustainable non-violent futures.

This specifically Māori orientation of including the whānau, was adopted in part, because the collective recognised that family violence is the product of processes of colonisation, the imposition of Western patriarchal ideologies, and issues of power and coercion (Krug, Dahlberg, Mercy, Zwi, & Lozano, 2002; King & Robertson, 2017). There is very little evidence to suggest that domestic violence was tolerated in pre-colonial Māori society (Durie, 2001). As with other indigenous cultures, there were behavioural norms, values and prescribed practices designed to moderate violence within our communities. However, these safety practices were thrown into chaos when Māori were forced to assimilate to colonising norms and regulations; were violently displaced from our ancestral homelands and communities; and were mustered into urban centres where domestic relationships take place behind closed doors and away from whānau scrutiny and protection (Masters-Awatere & Gosche, 2017).

Today, these interconnected factors serve as the complex backdrop to Māori having one of the highest rates of domestic violence in OECD countries. Māori are more than twice as likely to be the victim of intimate partner violence as any other ethnic group in Aotearoa/New Zealand. A particularly troubling aspect of domestic violence in Aotearoa/New Zealand is not only its prevalence among Māori, but also the rates of death due to intimate partner violence experienced by Māori. Māori women comprised 29% of those utilising refuge services nationally (relevant to our proportion of 15% in the general population) (Te Ara: The Encyclopaedia of New Zealand, 2018).

Through the history and development of Te Whakaruruhau, an entangled network of significant relationships developed in response to the disproportionate rates of domestic violence for Māori. These associations include the Māori & Psychology Research Unit (University of the Waikato), the Institute of Community Psychology, and more recently Ngā Pae o te Māramatanga.[3] Considering our 32-year history of collaboration, our collective engagements with Te Whakaruruhau can be described as multi-generational and enduring (Busch & Robertson, 1994; Hoeata, Nikora, Li, Young-Hauser, & Robertson, 2011; King, Hodgetts, et al., 2017; King, Rua, et al., 2017). Professors James and Jane Ritchie were initially involved from the beginning as part of their vision for a useful and engaged psychology embedded in communities and making meaningful contributions to improving peoples' lives. These Pākehā[4] scholars had a long career that included extensive experiences of residing alongside Māori communities, which shaped the psychology that they enacted. With colleagues, these pioneering scholars lead the establishment of Aotearoa/New Zealand's only community psychology programme (Robertson & Masters-Awatere, 2007) and the appointment of Māori staff to psychology at the University of Waikato from 1989, including Professor Linda Waimarie Nikora (Robertson & Masters-Awatere, 2007). Professor Nikora was taught by Professors Jane and James Ritchie, Professor David Thomas, and Dr. Neville Robertson who informed their community and research action with Māori cultural practices. These scholars created a space in the University for Māori Psychology to grow.

---

[3] New Zealand's Māori Centre of Research Excellence http://www.maramatanga.co.nz/.

[4] The Māori term for New Zealanders of British and European descent.

As a new appointee in 1989, Professor Nikora's brief was to grow a Māori Indigenous psychology and to support Māori students through psychology. After almost a decade in the academic discipline, Professor Nikora needed a mechanism to fast-track this objective. In 1997 she founded the Māori and Psychology Research Unit (MPRU) (Nikora, 2007), which became an important place for academics, researchers and community collaborators to meet, engage and strategise research and intervention activities with community groups like Te Whakaruruhau. Through the resulting dynamic and problem focused web of connectedness, arose opportunities for meeting our cultural obligations towards one another and contributing to the development of future generations of scholars. These new scholars obtained the skills to be useful to Māori communities generally from their upbringing as Māori and for working with agencies such as Te Whakaruruhau specifically also through their training in community psychology. As the original Director of the MPRU for 20 years, not only did Professor Nikora achieve her goal of embedding a Māori indigenous psychology into the curriculum, the MPRU helped graduate the largest cohort of Māori PhD and Masters Psychology students. Linda has since moved on to become a Professor of Indigenous Studies at the University of Auckland, and Co-Director of Ngā Pae o te Māramatanga (NPM), but remains an active associate of the MPRU and Te Whakaruruhau.

The Co-Directors of the MPRU are presently Dr. Mohi Rua and Dr. Bridgette Masters-Awatere, who are former students of the MPRU and staff members of the Community Psychology (CP) programme at the University of Waikato. Professor Darrin Hodgetts also worked in the MPRU for over a decade on various projects with Te Whakaruruhau, prior to taking up a Chair in societal psychology at Massey University. He remains a research associate of the MPRU. Dr. Shiloh Groot and Dr. Pita King came through the MPRU as students and remain research associates, whilst holding academic positions at the University of Auckland and Massey University respectively. Today, the MPRU effectively has three branches built upon this scholarly whakapapa (genealogy) with Bridgette Masters-Awatere and Mohi Rua at the University of Waikato, Linda Nikora and Shiloh Groot at the University of Auckland, and Darrin Hodgetts and Pita King at Massey University.

Through our multi-generational collaborations colleagues from Te Whakaruruhau and the MPRU have engaged in over a dozen specific research projects involving the co-design and evaluation of services as well as informing national policy developments and the work of aligned community groups and government agencies (Busch & Robertson, 1994; Hoeata et al., 2011; King, Hodgetts, et al., 2017; King, Rua, et al., 2017; King & Robertson, 2017; Masters-Awatere & Gosche, 2017; Ritchie, 1992; Robertson et al., 2013; Robins & Robertson, 2008). As important as these formal projects and reports are, what is even more important are the relationships that make such work possible and the informal cooperation that binds us in ongoing collaboration.

In referring to the importance of residing together, we often cite the mantra of 'relationships first and research second'. This mantra is reflected overtly in our ongoing collaborations with Te Whakaruruhau and knowing that we are there to sustain each other. This relationship has functioned beyond formal research projects since the time of the Ritchie's. Although particular projects are important, it is the relationships that enable these projects to be collaborative and co-designed that are the most important aspect of our working and being together. It is useful to think of particular projects as focused points of work within a broader flow of engagement. Sustaining these relationships requires us to trust each other, and be accountable to one another. This is helped by us having intimate knowledge of each other's work, the demands of our jobs, and the cultural principles that bind us as Māori. We also nestle upcoming generations of scholarly activists and agency staff (often the same persons). For example, our theses students are often the ones working daily on-site in Te Whakaruruhau, often being jointly supervised by one of us and an agency staff member who has previously come through the MPRU.

Our collective work with Te Whakaruruhau has required us to avoid getting ahead of ourselves as psychologists and to step out beyond our discipline as we prioritise the immediate needs of Māori communities. That being said, we are not as naive as to propose that we lack expertise. As scholarly activists we have capacities and resources that can be leveraged to support community-based work in gaining resources that increases the sustainability of such efforts. As such, we have become a part of Te Whakaruruhau's reflective practice and a mechanism for checking the efficacy of particular developments and strategies. A key lesson for community psychologists here is that we are often more effective when we are not in charge or developing and imposing (ineffective) interventions on communities. That is, our work often involves following the lead of agency staff rather than setting the agenda ourselves. We were prepared culturally to act in this way by observing obligations towards Māori notions of Whakaiti (service to others with humility), Hūmarie (acting with gentleness and kindness), and Kotahitanga (reaching unity through consensus). By doing so we get asked to be involved in a range of activities and can in turn call upon Te Whakaruruhau to do the same.

More specifically, Te Whakaruruhau invite members of the MPRU to sit in on important meetings, to help gather information in preparation for discussions with funders and to speak with them at public events. Members of the MPRU also invite staff from Te Whakaruruhau to come and speak to our classes, to co-supervise graduate students with us, to be advisors on projects with other community groups, and to collaborate on research that extend beyond the specific issue of family violence (King, Hodgetts, et al., 2017; King, Rua, et al., 2017). Doing so further entangles us. Rather than being two distinct groups—one community based and the other university based—who collaborate with each other, we function as the same group or network. This connection is strengthened by several of us holding the same tribal affiliations which demand cultural obligations to each other beyond the work setting. Staff from Te Whakaruruhau are also members of the MPRU, and we as the MPRU regularly carry out governance and advocacy work with Te Whakaruruhau.

Beyond the entangled relationship between Te Whakaruruhau and the MPRU that was until recently anchored in community psychology at the University of Waikato, we are now also affiliated with the broader national network of Māori scholars and community leaders that is Ngā Pae o te Māramatanga. The development of Ngā Pae o te Māramatanga under Professor Nikora's recent leadership has expanded the joint influence of the MPRU and Te Whakaruruhau into additional relationships with other community groups, scholars, policy makers, and government. To round out this section we will now recount two recent collaborative symposia designed to share knowledge and foster cooperation across efforts to improve the everyday lives of Māori. These symposia leverage off the broader network and resources that Ngā Pae o te Māramatanga affords us.

The first symposium was organised by Ngā Pae o te Māramatanga in partnership with central government to focus on how we strengthen responses to the problems that plague our communities. The second symposia was led by Te Whakaruruhau and was focused on the issue of family violence. Both events reflect an intensification of activity that has come with the recent change of government and their willingness to engage meaningfully with community in addressing social issues. Both symposia epitomise how our combined work is dynamic, engaged and responsive because it is occurring through a dialectics of community. Both symposia comprise culturally textured spaces in which self-advocacy from the community to government decision makers is centralized (*cf.*, Gierbolini, this collection). Both symposia also reflect a way of operating dynamically as a group of community trained Māori psychologists. It centralises the importance of being alert to developments and opportunities that arise in ways that our scholarly institutions and discipline are often painfully slow to recognise, let alone respond.

The first symposium, constituted a 1-day gathering jointly organised by Ngā Pae o Te Māramatanga and a central government research agency (SUPERU[5]). It was titled '*Toi Tū Te Whānau, Toi Tū Te Kāwai Whakapapa: A workshop on whānau and whakapapa for public policy*'. Our joint presentation was based on a 3 year Ngā Pae o Te Māramatanga funded project exploring the everyday lives of whānau from the Māori precariat (Rua, et al., 2019). We focused on links between government policy, growing rates of job insecurity and homelessness, and the inadequacies of penal welfare. We emphasized the need for culturally-patterned ways of engaging in research that generates pathways for action to address the colonial structures that continue to impoverish our people. By demonstrating our united front we were able to draw key decision makers into conversations about the impacts of current policies on the everyday lives of Māori. Central to the practical recommendations at the end of the presentation, was an account of the utility of the very core cultural concepts (Whānaungatanga, Manaakitanga, Kotahitanga, Whakaiti and Hūmarie) we have outlined in this chapter. We focused on how these Māori cultural concepts can be applied to the redesign of our labour laws, housing policies and welfare system. In engaging and hopefully corrupting members of our audience to join us, we worked to realise forced multiplier effects (Hall & Lamont, 2013) whereby we can achieve more for whānau by working together, than any one group can do alone. The session was highly successful and we have just completed a written policy 'Think-piece'

[5] The Social Policy Research and Evaluation Unit (SUPERU) was a government agency that focused on what works to improve the lives of families and whānau http://www.superu.govt.nz/.

(Rua, et al., 2019) outlining how we can address the needs of Māori communities and assist people in realising lives beyond poverty. The 'Think-piece' has now been distributed throughout central government.

This symposium also offered an opportunity for Te Whakaruruhau staff, along with several authors of this chapter, to present and engage directly with central government policy makers. Linking communities and policy makers is about leveraging Whānaungatanga (relationships) to make a difference. It is about being good allies (or better yet, accomplices) and cultivating support from those in power, including Māori and Pākehā (New Zealanders of European decent) who occupy key positions within government ministries. Our value to each other is the doors we can open when we transcend divisions between Western styled academies, community agencies and government. This is a form of community based scholarly activism where we work alongside the community to lobby for change. Such events enable Te Whakaruruhau to talk directly to those with the power and resources to help or hinder the aspirations of whānau living in poverty for a better life.

Following on from the joint NPM and SUPERU symposium to policy makers, was the '*Partnering for Whānau*' symposium organised by Te Whakaruruhau and funded by NPM. The 2-day symposium attracted presenters working for government organisations, community practitioners, iwi (Māori tribal groups) members, researchers, universities, as well as survivors and perpetrators of domestic violence. We had a chance to listen to ministry staff as well and to relate their agendas to our own work. More importantly however, the focus of Te Whakaruruhau was front and centre on this organisation's culturally patterned ways of working. In this forum opportunities for connectedness, engagement and exchange of ideas and practices as determined by the agenda of Te Whakaruruhau (and the needs of whānau accessing the service) were prioritised. The centrality of Te Whakaruruhau to such events has reached the point that they are now sort out by government for advice in the policy-making process.

In concluding this section it is important to acknowledge that while our approach is driven by core Māori cultural concepts, central are the contributions of our Pākehā accomplices who also 'speak out' and challenge hegemony, colonial structures and offer solutions to the issues faced by our communities. This can be liberating for Pākehā as well as Māori because non-Indigenous peoples are also the products of colonization and as such decolonisation needs to include Pākehā as well. For example, Dr Neville Robertson is a Pākehā man who has contributed to the efforts of Te Whakaruruhau in culturally informed ways (Robertson et al., 2013; Robins & Robertson, 2008) to the point that Te Whakaruruhau now refer to him affectionately as their 'Kaupapa Māori' researcher. Has been gifted with such a title because he recognises and accepts his accountabilities both as a community psychologist, as a member of the dominant colonial group, and we are also accountable to him. The key point here is that he has the cultural competence to accept that his supportive and action-orientated work with Māori communities is and should always be Māori-led.

Conclusion

Some forms of psychology can often be irrelevant, or slow to respond to community needs. When psychologists do act many behave as if they know what is best for the community. Part of the issue here is that as psychologists, we tend to be theory driven and many of our theories and methods of engagement are developed in very different contexts with little or no engagements with the communities with whom we end up working. Like many other authors of chapters in this book, we seek to avoid the top down theory to application process. Our communities cannot wait for the great minds of the academy to debate the intricacies of concepts such as epistemology. The alternative approach we adopt stems from the recognition of the need for Māori psychologists and allies/accomplices to respond to and serve community needs, ensure that Māori survive now and can live well into the future. Our approach is guided by the concepts of 'whakaiti' or humbling ourselves within collaborative relationships (whānaungatanga) that lead to acts of care (Manaakitanga) for whānau (immediate and extended family). This approach requires us to reside within relationships with community members, particularly those who are engaged in helping whānau in need. This work takes time, requires a team, and means that one may be engaged for decades. This chapter constitutes an effort to conceptualise what we have been doing in the hope that it resonates with colleagues engaged practically with similar concerns.

As recommended by contributors to this book (Gierbolini, this collection), we have foregrounded the importance of our cultural worldview and ways of being for how we relate to one another and our scholarship. In doing so, we have presented knowledge production as a culturally-textured and responsive process that is shaped by context and existing human relations both within and across ethnic groups. The common cultural ground we share with our community partners means that our bonds and obligations towards each other transcend colonial occupational categories such as professor and service worker and provide a basis for equitable and trusting inter-organisational relationships. The network within which we are immersed functions as a kind of insurgent movement for knowledge production and action in which we all contribute a piece of the puzzle of how to ensure healthy lives for Māori. As part of this process of leveraging our networks for collaborative action, is the production of research artefacts that bridge the community and policy (where the decisions that are made impact on communities) divide and which keep us all honest in terms of what is and is not working for Māori. As evidenced by this chapter, such work requires us to enact relationships that exist well beyond the discipline of psychology. It also requires us to sidestep our neoliberal institutions (Shore, 2010) that reward abstract theorising over collective responses to oppression.

What is particularly clear is that the days of community psychologists in Aotearoa/New Zealand turning up to a community to impose a preconceived plan on people, or to extract their knowledge, should be over. Many communities know what to expect and now prefer to engage their own researchers. In the last 18 years approximately 1000 Māori have graduated with PhDs and are now populating service providers, tribal research centres, wananga (Māori tertiary institutions) and government agencies. These graduates are central to processes of decoloniality. Many of the graduates from psychology actually prefer other workplaces, to schools of psychology, which remain dominated by WEIRD psychologies that have little or no relevance to the aspirations of their communities (*cf.*, Disrupting the canon chapter). Many prefer engaging in material change in society, over having to constantly justify their existence in a discipline that struggles to embrace diversity of perspectives. How we work with and deploy these networks to shape collaborative efforts to enhance peoples' lives as Māori is of pressing importance. Working collaboratively with such activists to realise our obligations towards intergenerational efforts to achieve the survival and thriving of Māori today and into the future is of primary concern. It is also a fulfilling way to spend one's time.

References

Busch, R., & Robertson, N. R. (1994). "Ain't no mountain high enough (to keep me from getting to you)" An analysis of the Hamilton Abuse Intervention Pilot Project. In J. Stubbs (Ed.), *Women, male violence and the law* (pp. 34–63). Sydney: Institute of Criminology.

Crenshaw, K. (1989). Demarginalizing the intersection of race and sex: A Black feminist critique of antidiscrimination doctrine, feminist theory and antiracist politics. *University of Chicago Legal Forum, 140*, 139–167.

Durie, M. (2001). *Mauri Ora: The dynamics of Māori health*. Auckland: Oxford University Press.

Enriquez, V. (1992). *From colonial to liberation psychology. The Phillippine experience*. Quezon City: University of the Phillippines Press.

Groot, S., Rua, M., Awatere-Masters, B., Dudgeon, P., & Garvey, D. (2012). Ignored no longer: Emerging indigenous researchers on indigenous psychologies. *Australian Community Psychologist, 24*(1), 5–11.

Hall, P., & Lamont, M. (2013). Why social relationships matter for politics and successful societies. *Annual Review of Political Science, 16*, 49–71.

Henrich, J., Heine, S. J., & Norenzayan, A. (2010). The weirdest people in the world? *Behavioral and Brain Sciences, 33*(2–3), 61–83.

Hill Collins, P. (2000). *Black feminist thought*. New York: Routledge.

Hoeata, C., Nikora, L. W., Li, W. W., Young-Hauser, A. M., & Robertson, N. (2011). Māori women and intimate partner violence: Some sociocultural influences. *MAI Review, 3*, 1–12. Retrieved from http://www.review.mai.ac.nz/index.php/MR/article/view/443

hooks, B. (1994). *Outlaw culture: Resisting representations*. New York: Routledge.

Hutchings, J., & Lee-Morgan, J. (2016). *Decolonisation in Aotearoa: Education, research and practice*. Wellington: New Zealand Council for Educational Research.

Jahoda, G. (2016). On the rise and decline of 'indigenous psychology'. *Culture & Psychology, 22*(2), 169–181.

King, P., Hodgetts, D., Rua, M., & Morgan, M. (2017). Disrupting being on an industrial scale: Towards a theorization of Māori ways-of-being. *Theory and Psychology, 27*, 725–740.

King, P., & Robertson, N. (2017). Māori men, relationships, and everyday practices: Towards broadening domestic violence research. *AlterNative: An International Journal of Indigenous Peoples, 13*(4), 210–217.

King, D., Rua, M., & Hodgetts, D. (2017). How Māori precariat families navigate social services. In S. Groot, C. van Ommen, B. Masters-Awatere, & N. Tassell-Matamua (Eds.), *Precarity: Uncertain, insecure and unequal lives in Aotearoa New Zealand* (pp. 124–135). Auckland, New Zealand: Massey University Press.

Krug, E. G., Dahlberg, L. L., Mercy, J. A., Zwi, A. B., & Lozano, R. (2002). *World report on violence and health*. Geneva: World Health Organisation.
Land, C. (2015). Reckoning with complicity. In *Decolonising solidarity*. London: Zed Books.
Masters-Awatere, B., & Gosche, J. (2017). Pasifika women affected by domestic violence: The case of Teuila. In S. Groot, C. van Ommen, B. Masters-Awatere, & N. Tassell-Matamua (Eds.), *Precarity: Uncertain, insecure and unequal lives in Aotearoa New Zealand* (pp. 203–215). Auckland, New Zealand: Massey University Press.
McGuinne, J. S. (2013, February 25). *Decolonisation*. Retrieved from https://johansandbergmcguinne.wordpress.com/2013/02/25/decolonisation/
Mead, H. (2003). *Tikanga Māori: Living by Māori values*. Wellington: Huia Publishers.
Ministry of Health. (2015). *Tatau Kahukura: Māori health chart book 2015* (third ed.). Wellington: Ministry of Health.
Nikora, L. W. (2007). Maori and psychology: Indigenous psychology in New Zealand. In A. Weatherall, M. Wilson, D. Harper, & J. McDowall (Eds.), *Psychology in Aotearoa/New Zealand* (pp. 80–85). Auckland, New Zealand: Pearson Education New Zealand.
Nikora, L. N., Masters-Awatere, B., Rua, M., Hopner, V., Alefaio, S., Stewart, L., et al. (2017). Indigenized internationalisation: Developments and lessons from two Aotearoa/New Zealand universities. In G. Rich, U. Gielen, & H. Takooshian (Eds.), *Internationalizing the teaching of psychology* (pp. 129–142). Charlotte, NC: Information Age Publishing.
Ritchie, J. (1992). *Becoming bicultural*. Wellington: Huia Publishers.
Robertson, N., Masters, B., Lane, C., Tapara, A., Corbett, C., Graham, R., et al. (2013). *Evaluation of the Whānau Ora wellbeing service of Te Whakaruruhau: Final report*. Hamilton, New Zealand: University of Waikato, Māori and Psychology Research Unit.
Robertson, N., & Masters-Awatere, B. (2007). Community psychology in Aotearoa/New Zealand: Me tiro whakamuri a kia hangai whakamua. In S. Reich, M. Riemer, I. Prilleltensky, & M. Montero (Eds.), *The history and theories of community psychology: An international perspective* (pp. 140–163). New York: Kluwer Academic Publishers.
Robins, K., & Robertson, N. (2008). *Te Whakaruruhau transition and wellbeing programme: An implementation evaluation*. Report prepared for Te Whakaruruhau and Te Puni Kokiri. Hamilton, New Zealand: Māori and Psychology Research Unit, University of Waikato.
Rua, M., Hodgetts, D., & Stolte, O. (2017). Māori men: An indigenous psychological perspective on the interconnected self. *New Zealand Journal of Psychology, 46*(3), 55–63.
Rua, M., Hodgetts, D., Stolte, O., King, D., Cochrane, B., Stubbs, T., . . . Groot, S. (2019). Precariat Māori Households today: the need to reorient policy to cultivate more humane understandings of whānau in need. *Ngā Pae o Te Māramatanga Te Arotahi Series(May)*, 1–16.
Seedat, M. (1997). The quest for liberatory psychology. *South African Journal of Psychology, 27*(4), 261.
Shore, C. (2010). Beyond the multiversity: Neoliberalism and the rise of the schizophrenic university. *Social Anthropology, 18*(1), 15–29.
Smith, L. T. (1999). *Decolonising methodologies: Researching & indigenous peoples*. Dunedin, New Zealand: University of Otago Press.
Sonn, C. C., Rua, M., & Quayle, A. (2019). Decolonising applied social psychology: Culture, indigeneity and coloniality. In K. O'Doherty & D. Hodgetts (Eds.), *Handbook of applied social psychology*. Thousand Oaks, CA: Sage.
Te Ara: The Encyclopaedia of New Zealand. (2018). *Defining violence*. Retrieved from https://teara.govt.nz/en/domestic-violence/page-1
Te Awekotuku, N. (1991). *Mana wahine Maori: selected writings on Māori women's art, culture, and politics*. Auckland: New Women's Press.
Tuck, E., & Yang, K. W. (2012). Decolonization is not a Metaphor. *Decolonization: Indigeneity, Education & Society, 1*(1), 1–40.
Waitoki, W. W., & Levy, M. (2016). *Te Manu Kai i Te Mātauranga: Indigenous psychology in Aotearoa/New Zealand*. Wellington, New Zealand: New Zealand Psychological Society.
Walia, H. (2013). *Undoing border imperialism*. Oakland, CA: AK Press/Institute for Anarchist Studies.
Watkins, M., & Shulman, H. (2008). *Toward psychologies of liberation*. Basingstoke, UK: Palgrave Macmillan.

Disrupting the Psychology Canon? Exploring African-Centered Decolonial Pedagogy

Hugo Canham, Lesiba Baloyi, and Puleng Segalo

We begin by defining and outlining the problem of the Western canon and its hold over the discipline of psychology in the South African university. We locate ourselves as 'decolonial dreamers' by providing three experiential accounts that inform our standpoints in relation to our attempts to disrupt through foregrounding African narratives and classroom practices. Finally, we plot a path for how this canon can be disrupted by offering an alternative community psychology that is grounded in African epistemologies.

As black people, we are conscious of the fact that the university was not built with us in mind. This makes our role as academics both ironic and precarious. As a group of authors, we were deliberate in partnering for this chapter. We wanted to record the experiences of black people with similar commitments to decolonisation at three different universities. In sharing our experiences and methodologies, we sought to learn from each other and to commune in supportive fellowship. We therefore share our experiences in order to illuminate the challenges and possibilities for crafting an African-centred decolonial pedagogy. We do this by presenting the value of a contextually affirming psychology and the emotional burden for disrupting a long established canon. Here, we (three authors) take turns to give first person accounts of our pedagogical encounters with students in the classroom. Our experiential accounts are written in order to explicate the theoretical arguments that we make in this paper. For example, we demonstrate the recalcitrance of the canon

H. Canham (✉)
School of Human and Community Development, University of the Witwatersrand, Johannesburg, South Africa
e-mail: Hugo.canham@wits.ac.za

L. Baloyi
School of Medicine, Sefako Makgatho Health Sciences University, Pretoria, South Africa

P. Segalo
College of Human Sciences, University of South Africa, Pretoria, South Africa

G. Stevens, C. C. Sonn (eds.), *Decoloniality and Epistemic Justice in Contemporary Community Psychology*, Community Psychology,
https://doi.org/10.1007/978-3-030-72220-3_11

and how we attempt to disrupt Western epistemologies and the attendant challenges that we encounter, and we offer African centered alternatives to community psychology crafted around the philosophy of Ubuntu. We do this by highlighting how our own narratives simultaneously disrupt coloniality and enrich our aspirations for contextualism. The accounts represent our different personal histories, experiences and voices and illustrate some of the principles we have highlighted in this paper. We demonstrate how we bring our bodies, histories and communities to the pedagogical encounter. We aim to create opportunities for learning engagements which are important in knowledge creation.

The Western Psychology Canon in the African Academy

Canons or canonical thought represent well-trod ideas around which there is some consensus in relation to legibility, representation, validity, ontological truth claims, and ways of knowing these truths. This is to say that there are canonical methodologies and interpretations that are backed by institutionalisation, scholarly and published work, the acceptance of a single standard, and a transhistorical universalistic value of what constitutes the exemplary. Reading lists and material that circulate in mainstream university spaces are generally representative of Euro-American canonical thoughts and practices. The canon is not neutral, and neither is it the best about what is known in the world (Marx, 2004). It is synonymous with the Westernised university. For Grosfoguel, Oso, and Christou (2015), this university is any university in the world that has its roots and knowledge foundations in the Western canon of thought. Grosfoguel et al. (2015, p. 74) contends:

> Our job in the Westernized university is basically reduced to that of learning these theories born from the experience and problems of a particular region of the world (five countries in Western Europe) with its own particular time/space dimensions and "applying" them to other geographical locations even if the experience and time/space of the former are quite different from the latter.

Pollock (1999, p. xiii) proposes that the canon can be understood as simultaneously discursive in structure and as "masculine narcissism within the exercise of cultural hegemony". Freud's psychoanalytic theory and Piaget's theory of development are prominent examples of culturally hegemonic approaches based on masculine narcissism. Pollock, however, omits the racist narcissism that informs Western canonical forms of thought. Franz Fanon (1967) was strident in his criticism of how the black man's (sic) psyche was constituted by the colonial situation. Similarly, Manganyi (1973) argued that the oppressive conditions of apartheid that black people lived under had created a personality that reflected these conditions. For both these African theorists, it was clear that context and culture mattered for understanding the human subject. These scholars read against the grain of the canon by centering the location of their theorisation. The question of what constitutes truth, reality, or what is known cannot therefore be understood from a universalist point of view.

The Euro-American canon was created and consolidated during the relatively undisrupted centuries of knowledge production in Europe and North America. This canon was founded on extractive economies during the colonial period. Africa was used as a petri dish to enable the study of disease. Africans were enslaved for free labour to contribute to the Western canon in the construction of infrastructure in the worlds burgeoning metropoles (Sharpe, 2016). Music, art, and raw materials were harvested from the continent and other colonial territories to bolster the Western canon. The pillaging of resources meant that African knowledges were actively disrupted and denigrated by colonial settler communities, missionaries and colonial administrators. In the geopolitics of knowing, colonialism saw the colonies as having culture (and data), and Euro-Americans as having science (Mignolo, 2009).

Despite the Euro-American subjugation of Africans and dominance over indigenous knowledges, African knowledge systems survived and developed over time. The use of African materials, labour and data, on the continent and elsewhere, were central in the production of knowledge and the Western canon. The intervention that we offer here then is a claim that Africa features and is co-constitutive of world knowledges. What is at stake however, is the positioning of Africans as objects in relation to these knowledges, rather than as producers of authentic knowledge. The decolonial project is therefore one of reclamation, reassessing what constitutes knowledge, and retraining the gaze (Mignolo, 2009). This project is also about mounting a critique of the ways in which the Western gaze makes objects of those on its margins (Painter, Terre Blanche, & Henderson, 2006) while also creating something new (Mignolo, 2009).

The transition from the apartheid system in 1994 to a democratic government in South Africa did not transform the apartheid education system. The changes that are claimed to have occurred in higher education have failed to address the thorny and critical issue of epistemological and curriculum transformation. African cultural practices, experiences, behaviours, performance of rituals and other ways of knowing and doing continue to be viewed and interpreted using foreign Western epistemologies. These ways of knowing derogate local knowledges. We argue that these changes amount to cosmetic disguises meant to maintain the conquerors' pedagogical system of thought. Disrupting Western canons of scholarship requires alternatives to the single narrative that infuses current epistemological, methodological and teaching approaches. The alternative must include knowledge systems that value multiple realities. This arises from the view that the Western knowledge system is but one of the many knowledge systems in the world and cannot therefore claim monopoly of knowledge production (Mignolo, 2009; Mkhize, 2004).

There is some consensus that the teaching and practice of psychology in South Africa (SA) is almost wholly informed by the Western canon (Ramoupi, 2014; Ratele, 2015). Writing more broadly about the epistemological underpinnings of knowledge in SA, Neo Lekgotla Laga Ramoupi (2014) noted that the theoretical position and curriculum was largely unchanged and generally unquestioned since the demise of apartheid. With reference to psychology, Kopano Ratele (2015) contends that African, anti-colonial, postcolonial, and de-colonial social thought are absent from the curriculum. He critiques our failure to decolonise the curriculum: "our higher education system, the curriculum we design, classes we teach, and research we generate, continue to reproduce, or simply adjust, values supportive of racist, hetero-patriarchal, fundamentalist-capitalism" (2015, np). More recently, Ratele (2017) has argued that European epistemologies, values and interests, continue to inform the basis of South African psychology.

Decolonial Praxis and Disrupting the Canon

The discipline of psychology, like many other disciplines continues to suffer from what Lewis Gordon (2006) calls *disciplinary decadence* where foundational Euro-Western theories and theorists are revered and alternative ways of knowing are eschewed. This leaves little room for questioning, critiquing and reshaping what psychology is. The colonial lens that continues to be used to make meaning of life and psychological challenges can be perceived as a form of epistemological violence (Teo, 2010) that assumes the lack of history and existing knowledge outside of the statistical diagnosis. For example, the use of the Diagnostic Statistical Manual of Mental Disorders as a universal standard for diagnosing, understanding and treating pathology does not account for lives outside of the Euro-American frame. We argue that this is a form of coloniality. Our understanding of coloniality follows Andreotti, Stein, Ahenakew, and Hunt (2015, p. 23) who contend that it "can be understood as a system that defines the organization and dissemination of epistemic, material, and aesthetic resources in ways that reproduce modernity's imperial project". Coloniality is inside our psychology textbooks and within the walls of colonial universities that many of us occupy. We therefore call for the decolonisation of the teaching of psychology in general, and of community psychology in particular. To decolonise is to deal with the oppressive structures within which we exist and operate in the wake of colonialism.

Grosfoguel et al. (2015, p. 647) point us to *critical decolonial thinking* which they describe as "the epistemology that emerges in colonial situations where the hegemonic perspective is subverted from the cosmologies, languages and epistemologies of the subaltern". They contend that this is a new form of epistemology that emerges in the "in-betweenness of two languages, two cosmologies, two epistemologies, where the subversion of the hegemonic knowledge is produced from the geopolitics of knowledge of the subaltern". For us then, the challenge is to disrupt by inserting ourselves while refashioning hegemonic knowledges in ways that serve us. This position enables us to be anti-Eurocentric without being anti-European. While appreciating the contributions made by European theories, we argue for the right to critique the ways in which they have subordinated other epistemologies. We contend that there must be a space for dialogical engagement between Eurocentric and African knowledge systems on an equal basis.

Westernised universities are structures of power that continuously reproduce themselves by claiming the sole right to define and determine the nature of truth and what can be taught. In order to problematize this hegemony, there is a need for epistemic disobedience (Mignolo, 2009) where we are not restricted by pre-determined ways of knowing. This is necessary for the creation of multi-epistemic spaces that will facilitate dialogical engagements. Multi-epistemic spaces should represent and be reflective of a confluence of different sites of knowledge systems. The curriculum is one of the spaces where hegemonic practices should give way to multiple ways of knowing. Properly construed, decolonising the curriculum has to create an education system that conscientises people by infusing critical thinking in teaching and learning practices (Freire, 1972). The purpose of education is to transform society. Epistemological imprisonment of alternative forms of viewing the world can have debilitating effects that can be likened to a cancerous growth. This cancer eats away at people, slowly mutating and metamorphosing from within and manifesting externally in the continuous injustices and inequalities that permeate higher education. The process of decolonisation seeks to provide tools for possible treatment of the cancer so that higher education produces 'healthy', responsible, and justice seeking citizens. To illustrate the lived endeavours and challenges of working alongside and against the Western canon, we provide an experiential account by the first author.

Disruption, Resistance and Stalling

I (the first author) work at one of the largest departments of psychology in the country. With heightened pressure by students to decolonise the curriculum, the department decided to appoint a "senior" member of faculty to lead and advise on curriculum *transformation*. The word "decolonisation" was elided in this naming and muted in subsequent conversations. It bobs like a cork on oily water. I was tasked with the role of thought leader on curriculum transformation. In this reflection, I note two events that epitomise my experience within this role. I centre critical incidents in order to understand the pattern of events within institutional contexts. I begin in the classroom by thinking through an experience that occurred at the time of writing this piece. I was teaching qualitative methods to an honours level postgraduate class. It was the first time that I had taught this course and I had deliberately framed it around critical methodologies informed by decolonial methods and community contexts. I emphasised the philosophical underpinnings of qualitative methods. The course generated heightened discussion. But I also noted that some students had become snappy, irate, and sullen. In the sixth seminar, one of the sullen students snapped and launched into a critical, if somewhat cathartic, monologue in critique of the course. "I can't say that I have learned anything about methods in this course. We have spoken about everything but methods! It is more like a philosophy course. A lot of unfocussed talk." A debate ensued but the student was steadfast in her argument that this course had not taught any technical methodological skill. She suggested that there were a number of other students that agreed with her.

In the moment, I was convinced that this was an instance of marginal knowledges rubbing up against conservative mainstream knowledges that had been developed over a period of time. Feminist and queer methodologies and African philosophical ways of seeing had unsettled and threatened some students' sense of self. In part, I had known that the curriculum would create discomfort. Following Freire (1972) and Pinar (2012), discomfort is an important part of the pedagogic process. But the discomfort can also threaten the privilege of those who are comfortable in the classroom. For me, this recalls other experiences of teaching community psychology where students' discomfort with the curriculum led to heightened alienation for some students and conflict in the classroom. White students have seen themselves mirrored in the Western canon whereas black students in psychology have a history of alienation from the curriculum (Kiguwa & Canham, 2010). The attempt to centre the black experience troubles the Western canon. The outburst can therefore be read as an attempt to resist the ways in which the black and feminist African woman's body had become the point of reference in the course. In this instance, I resisted the push to recentre the white experience and its valorisation in the Western canon. For me, teaching research methods and community psychology for the uncritical extraction of black experience perpetuates epistemic violence. The black women in the qualitative methods class were largely quiet but they talked to me beyond the confines of the classroom. Unaccustomed to being at the centre of knowledge, they were appreciative but curious about my interest in black women. This questions my own positionality as a black man in the pedagogic encounter.

In some sense then, I have some agency and space to disrupt the Western canon in the classroom. However, the remaining scenarios illustrate that I have less sway in my role as curriculum transformation thought leader in my department. At stake is what we as academics consider as knowledge and the implications thereof. In the following examples, I illustrate institutional resistance and how it stalls and stifles possibilities for change even as it retains curriculum transformation as a running agenda item for meetings.

The coordinator of an undergraduate course has called a meeting. She, a black woman, and her colleague, a white man, who will teach the same course, have reached a stalemate in relation to the changes that they will make to the curriculum. They need to come to a decision because the course will commence soon. I have been called into the meeting to assist in providing some conceptual clarity in relation to how the department of psychology envisages curriculum transformation. There are a number of issues at play here: curriculum transformation versus decolonisation of the curriculum, academic freedom to determine curriculum, African writing as foundational theory or as critique, how racialised bodies are implicated and placed in relation to disruption and maintenance of the status quo, and fear of change. Some of those in the meeting mount a well-considered argument that African writers such as Fanon are theorists who go beyond criticism of the Western canon to create theory of the black psyche. There is contestation as to what constitutes a core curriculum and suggestions to defer "challenging" content to later years of study. The effect of the postponements is that, besides cosmetic changes, the curriculum remains the same at the undergraduate level. Meetings, strategies and plans for curriculum transformation that amount to a performance, can be seen as ways of masking colonial lineages. Maldonado-Torres (2007) has termed this coloniality. To be sure, while the meeting is more than a performative gesture, it will end in compromise and deferments. Working relationships have to be considered.

The black woman who has called this meeting is left with the residue of the bad feelings that come from mounting difficult conversations. Sarah Ahmed (2009) points to the double bind that confronts black women in academia. They embody diversity through their raced and gendered identities, and they are expected to bring about change. For these reasons, they are seen as the provocateurs. The consequence is that they attract resistance and ill feeling. They are seen as the causes of unhappiness. Woodard and Mastin (2005) have illustrated the ways in which black women who work beyond the bounds of housework are described as aggressive, outspoken and controlling (Canham, 2014). In this instance, the woman tells me that it has taken a week to shake off the bad feelings that remained with her from the meeting. But the residue remains with her. She is caught between teaching her version of a decolonised curriculum and avoiding a prolonged collision and more bad feelings.

The preceding scenario has become commonplace as psychology departments attempt to engage with the realities of the moment. In other moments, when I have asked colleagues to consider what an African perspective is and what it would mean for their area of teaching, I have been told that the there are no African perspectives on the brain for example. "Because the brain is the brain is the brain!" The implication is a universalising discourse that sees cognition as one dimensional and Western. In this conception, Mignolo (2009, p. 1) contends, "The knowing subject maps the world and its problems, classifies people and projects into what is good for them". The possibilities of knowing from the South are foreclosed. The consequence of these encounters and others is that individual scholars are left to decide to teach what they like. At the institutional level, disruption is stalled. At best, there are pockets of change in the classroom encounter where some of us engage in the disruptive process of epistemic disobedience (Mignolo, 2009) in our efforts to counter the dark side of the shine of modernity (Andreotti et al., 2015).

Having sketched our criticism of the work of the Euro-American psychological canon and its deleterious ongoing effects, we now turn our attention to how coloniality affects community psychology and we offer an alternative in the form of an African-centred community psychology.

Coloniality and Community Psychology

The genesis of community psychology as an academic sub-discipline is contested (Fryer, 2008). For instance, Snow, Grady, and Goyette-Ewing (2000) and Toro (2005) identify the Swampscott Conference held in Boston, USA as the beginning of community psychology. Fryer (2008) traces it back to Jahoda's work in Marienthal, Austria in the 1930s. While South America, Asia and Africa do not claim origins, communal living is historically valorised and the contextual factors are foregrounded as central for the development of the discipline on these continents. We contend that there is no need for Africans to argue about the origins of community psychology, as this would be akin to questioning Africans' very existence and orientation to life (Baloyi & Ramose, 2016; Gyekye, 2002; Ramose, 2002).

In SA, Anonymous (1986) and Naidoo (2000) posit that the sub-discipline emerged as a response to the inadequacies of clinical psychology as practiced during apartheid. SA community psychology textbooks (e.g. Duncan, 2013; Duncan, Bowman, Naidoo, Pillay, & Roos, 2008) and a range of foundational writings in the field (e.g. Lazarus, 1988; Lazarus & Seedat, 1995) have aligned their guiding values and principles to "universal" values. For instance, Lazarus and Seedat (1995) identify four core purposes for SA community psychology: the extension of mental health services to all who need them, particularly underserved communities; the transformation of the ways that aetiology and development of psychosocial challenges are conceptualised; centering a contextual analysis that foregrounds social issues in order to transform the praxis of services rendered to include initiatives of prevention and resilience strengthening among vulnerable and high risk communities; and embracing advocacy, lobbying, community networking and mobilisation, and policy building. While these developments attempt to be aligned to local contexts, they lack the particularity that recognises and takes African worldviews seriously. For us, part of the decolonial impulse is to give texture to how community psychology advances local lives embedded in African cosmology. Here, we contend that South African community psychology should be more closely aligned with proponents of African psychology (Mkhize, 2004). Borrowing from African epistemologies, next we present the central tenets of a culture bound community psychology.

Recognising the nuanced differences between decoloniality, decolonisation, and Africanisation, we advocate for the recovery and centering of African epistemologies in ways that draw on the conceptual basis of all three perspectives. Briefly stated, we follow Tuck and Yang (2012) who contend that decolonisation is not a metaphor for cosmetic change but rather an act to return indigenous land and life. We view decoloniality as a related concept that is concerned with power in the creation of "radically distinct social imaginaries, conditions, and relations of power, knowledge" (Walsh, 2012, p. 11). To Africanise is to give African content and form to decoloniality and decolonisation. African land, life and knowledges are therefore centred in a decolonised African university.

African-Centred Alternatives to Community Psychology

There are multiple epistemological paradigms in the knowledge basket and these realities should be reflective in psychology as a meaning-making social inquiry enterprise. Despite calls for curriculum transformation, Western thought still dominates the academic philosophical landscape (Odora Hoppers, 2002). The colonial epistemologies that have introduced and sustained colonial thinking, culture and language in our academy must be challenged to create an African epistemology and psychology which will be representative and reflective of the diverse African voices in meaning, content, scope and structure of the psychology curriculum in general and the teaching of community psychology in particular. Nwoye (2018, p. 46) supports this view by stating:

> With its now clearly recognized inadequacies for the study of the life of the mind in Africa, traditional (mainstream) Western psychology can no longer be allowed to continue unchallenged in its hegemonic way of dominating the space for the study of psychology in African universities.

In our view, the need to challenge this Western hegemonic discourse and create a truly African psychology is based on the fundamental epistemological and methodological difference between Western and indigenous thought (de Sousa Santos, 2014; Kovach, 2010b). This difference causes philosophical, epistemological, cultural and methodological conflict for indigenous researchers and psychologists (Kovach, 2010a). Our insistence on the significance and meaningfulness of an African psychology community-based knowledge production is both necessary and imperative in creating the context for questioning the universality and applicability of Western educational epistemologies, methodologies and pedagogy in pursuit of epistemic and social justice in Africa. The African psychology curriculum we propose should be for and by African people, using theories, methods, sources and techniques that are reflective of the African worldview. The current psychology curriculum is based on the conqueror's philosophies, theories and methodologies which continue to privilege Western ways of knowing while denying the validity of indigenous knowledge, culture and languages (Makalela, 2017; Ramose, 2016; Ramose & Baloyi, n.d.; Smith, 2012). This exclusive deliberate focus on Western conceptions of knowledge production precludes the possibility of dialogical engagement or mutual coexistence between African and traditional Western psychology. The prevailing circumstances in psychology constrain African knowledge systems and experience both at an epistemological and ontological level. This means that it becomes unable to explore the multiplicity and complexities of the visible and spiritual human experience, as experienced in the African world (Nwoye, 2018). Our argument is that the exclusion, rejection or continued marginalisation of African knowledge systems from the psychology curriculum in the academy constitutes an ethical and social injustice and ought to be challenged (de Sousa Santos, 2014; Ramose, 2016). Imposing an alien community psychology curriculum that does not reflect the African worldview, reality and ethic on Africans constitutes an epistemic injustice. To be consistent with the African worldview and ethic, community psychology in Africa and South Africa should be based on the community model (Bujo, 1998).

Ubuntu and African Conceptions of Community and Personhood

In this section, we argue that the concept of community and, subsequently, community psychology in Africa, should be anchored on the African philosophical paradigm of ubuntu. According to Ramose (2005, p. 40) "the epistemology of *ubuntu* presupposes the ex-istence of *umu-ntu*, the human being". Since *ubuntu* demands *of abantu/batho*, people to relate in a respectful, humane manner, *abantu,* human beings are ontologically oriented towards the betterment of human relations and the good of humanity in the context of community. The foundation of community building is peaceful interpersonal relationships. Relationships are indispensable in the formation, sustainability and wellbeing of community. The African-ness of African people finds expression through the epistemology of *ubuntu* (Ramose, 2002). The glaring exclusion of *ubuntu* in many community psychology prescribed textbooks and training programmes constitutes epistemicide (de Sousa Santos, 2014) because, for Africans, *ubuntu* is an ethic of being human. Ontologically, the African conception of being human is communal in nature, and personhood is reflected in people's relationships with others and their cultural context (Bujo, 1998; Mkhizc, 2004). On this rcasoning, the inception of any African community psychology course must take this reality as its starting point. We now locate *motho/umuntu*, a person's identity and meaning in community.

For Black Africa, one's identity carries significant meaning to a person's ontology. Ontology is the philosophy of being whose manifestation is through human experience in given contexts (Ramose, 2016). The identity and being of an African is best understood in relation to other human beings within a family or community context. This is conceived as be-ing as a wholeness. A person's identity bears resemblance to their family of origin, and is inextricably linked to the community from which he or she comes. A person is born into an existing family, and a family is located within a community (Bujo, 1998). From an African worldview, *motho/umuntu,* a person is seen as an interdependent and interrelated culture-bound communal being. This is so because, from an African epistemological paradigm, personhood is defined in terms of wholeness. This does not deny human individuality as an ontological reality but ascribes ontological primacy to the community through which the individual comes to know and understand themselves and the world around them (Ramose, 2002). On this basis Ramose (2002, p. 64) argues:

> Wholeness is therefore the starting point of the African concept of a person. Consequently, the human person in African thought is not definable in terms of a single physical or psychological characteristic to the exclusion of everything else.

In the context of this discussion, the African worldview and experience should serve as the centre of any decolonising discourse in our universities. African people conceive the be-ing of a person as an integrated whole, unfolding in the "We" epistemology, contrary to the Western individualistic "I" manifestation of the person. It is this "We" epistemology that institutions of higher learning, and by extension the students they produce, have to orientate themselves towards in order to study and understand the African conception of a person's identity in community. Universities and students in these institutions have to resemble and serve the community within which they are located. Practically, this means that the curriculum taught at our institutions of education must reflect the cultural imperatives, developmental goals, needs and aspirations of the community they serve. The design of the psychology curriculum must be based on these socio-cultural patterns and features of the communities concerned. In this way, students of community psychology and subsequently graduates from universities will see themselves as active participants in the development and wellbeing of their communities. In our view, a decolonised curriculum must take this socio-ethical dimension of being a member of a community as its point of departure. According to Gyekye (2002, p. 299):

> Communitarianism or communalism therefore sees the human person as an intrinsically communal being, embedded in a context of social relationships and interdependence, never as an isolated, atomic individual... the notion of common interests, goals and values is crucial to an adequate conception of community, that notion in fact defines community... it is the notion of interests, goals and values that differentiates community from a mere association of individual persons.

Personhood therefore finds expression and meaning in the context of family and community through active participation. The participation of the individual is for the collective good of the community and humanity generally. Individuality is therefore not denied in African thought. Instead, as Gyekye (2002) argues, the community provides the cultural, social and contextual space for the attainment of individual desires and personhood. Additionally, Montero (2007) asserts that the "I" and the "other" are constructed inter-relationally and therefore construction of knowledge is always in relation. Reyes Cruz and Sonn (2011) contend that an epistemology of relatedness in indigenous writing in contexts such as South America and Australia and New Zealand, challenge Western conceptions of wellbeing by offering an indigenous ontology that emphasizes connectedness to land and the other. Looked at in this way, personhood is never independent of influence from others in the world. In the traditional African thought of *motho/umuntu,* personhood or to be truly human is always relational and communal and manifest in the philosophical ethic of *ubuntu*. We thus contend that decolonising psychology and community psychology curricula in the academy must be based on this communal understanding of what it means to be human in African thought. This would include an ethic of care and consciousness as argued by Montero (2007) who highlights that a liberatory (and decolonised) psychology has at its core an ethical aspect that acknowledges that liberation is a "constant battle, and critique, democracy, and participation are means to achieve it" (p. 529).

Here, we insert the experiential account of the third author to demonstrate alternatives to mainstream pedagogy. The third author records her own experience as a student of psychology and how this has influenced her teaching practices as an academic.

Redefining the Classroom/Learning Space

Schools and universities repeatedly assume that those who come in do so with the intention to fill their 'empty vessels'. In that process one's being and history is often erased. wa Thiong'o (1993) says this is akin to *new software* getting inserted. As an undergraduate community psychology student, as a black young woman, I (third author) had to 'go into communities' to offer my services and do practical training. The community (often a black community in a township) was presented as something that is 'out there' and needing to be rescued and assisted. This was the same community that I grew up in. Now, an academic institution was *forcing* me to re-enter as an outsider/alien and start using a lens that makes me perceive my community as lacking, vulnerable, marginalised, helpless and backward. With community psychology principles in the pocket, we had to use formulas to resolve the challenges faced by people in the communities. We had the power to make decisions on programmes to be put in place. This is how I became 'disciplined' into using boxes and categories to engage peoples' lived realities. These lenses were carried over into my academic career. In this process, I, and others like me become academics that consume, speak and defend the system; and we contribute in 'thingifying' ourselves (black people), and others like us. It was from the very same well that I too started drawing and I expected my students to do the same. However, the discomfort I felt began to surface and I had to confront the ways in which I was contributing to epistemic violence that is widespread within our academic institutions.

Through introspection (and in many ways fueled by the call to decolonise universities), I came to the realisation that teaching what I should is a matter of justice and ethics. Forcing students to leave their histories behind or at the door of the lecture hall is unethical. I began to confront and to fully acknowledge that the borrowed concepts that we have given a home in our minds cannot be used as a basis to imagine and transcend coloniality. By coloniality, I refer to the ways in which transformation (more black faces in front of lecture halls imparting knowledge) has taken place while the status quo (Eurocentric texts being relied upon) remains intact. Nyamnjoh (2005), citing Crossman and Devisch (1999, p. 11) asserts that "from independence to date, African universities have been successful in Africanising their personnel but not their curricula or pedagogical structures to any real extent… Education in Africa is like a pilgrimage to the Kilimanjaro of western intellectual ideals". The curriculum disempowers students and they end up existing outside of who they are supposed to be. There is a process of the remaking of an African (black) child at the point of encounter with higher education institutions. Many are left with internal dissonance as they are made to practice and theorise in ways that further dehumanises them, their families, histories and their communities.

It is at this juncture that epistemic rebellion becomes an imperative. There is a need to de-centre that which has been centralised, that is, Euro-American knowledge. This makes the need for decolonisation critical. Decolonisation must be about confronting the coloniality of knowledge in the classroom, and in dialogues among and between students and lecturers where space for critical engagement is allowed. This requires disrupting, delinking, being disobedient, and un-learning in order to re-learn. In my attempt to disrupt and move towards decolonising how I teach community psychology, I embarked on a conscious process of facilitating learning from outside the university walls. My efforts were fuelled by Freire's (1998, p. 93) assertion that "I cannot be a teacher if I do not perceive with even greater clarity that my practice demands of me a definition of where I stand".

I reflect here on a section of community psychology that I facilitate for the Master's in Research Psychology course called *Divergent Perspectives in Community Psychology.* The aim of the module is to provide an understanding of the many ways in which community can be understood. Furthermore, it aims to engage and trouble where and how learning takes place. At the Master's level, students are expected to show that they are independent thinkers who are able to contribute towards knowledge creation. I take students on a historical journey by literally taking them out of the classroom and the university premises. We have sessions at Freedom Park (one of the cultural heritage museum sites that tells the narrative of South Africa from creation through to colonisation, slavery, apartheid, and ultimately the attainment of democracy). Here, students are confronted with the history of South Africa that engages how black people were dispossessed of land, their cultures, religions, etc. They are afforded the opportunity to come face to face with the often uncomfortable aspects of the country's past and learn about how colonialism and the apartheid regime contributed towards people's current lives. It is hoped that this will challenge them to look at their research projects more broadly and challenge some of the taken for granted knowledges found in many of their textbooks. This process of journeying into the past is usually very challenging for students as it forces them to confront and learn about some aspects of their history that they were not aware of. They delve into uncomfortable spaces and initially (for some) there is some resentment and the need to be defensive. The walkabout is followed by reflection sessions where we focus on emotions that are evoked by going through this journey. We then link the process with the implications for the research work we do. We engage with what happens when history is told in distorted ways where certain aspects of the past are *hidden* and not found within academic spaces. I ask questions that are linked to knowledge production as it relates to our discipline of psychology and the communities 'out there'. I ask: 'For whom do we speak? With what authority do we speak? What do we know about the people we claim to represent? What is the role of psychology in history'? These are some of the questions we tackle in the quest to reimagine the role and relevance of psychology within communities. These engagements are backed by references to African

Psychology scholars' works that I seek to insert into conversation with the students' walkabout experience.

Family, Community and Questions of Land in African Community Psychology

African conceptions of family which includes grandparents, father, mother, children, cousins, aunts, uncles (visible beings) and the community of the living-dead and the yet-to-be-born (invisible beings), differs significantly from the Western family which consists of father, mother and children (Bujo, 1998). The African family and community thus consist of the visible and invisible or spiritual ontology. This extended family unity is crucial in understanding the African conception of family, its functioning and relationship to the community and the land. Those who passed on, *badimo/abaphansi*, the departed or living dead, continue to influence the actions of the living (Mkhize, 2004). After transitioning, the departed transcend to the spiritual ontology and are "rested", buried on land where they assume residence in the warm belly of motherland. However, the community of the living dead is never forgotten (Bujo, 1998). The land on which the transitioned are buried becomes an identity marker for the living and is regarded as the spiritual home for the living dead. Reconnection with the family is through communing with the living dead/ ancestral spiritual land. The spiritual connection between the living and the living dead serves as an affirmation of the integration of the living, the living dead and the land. The very being of African people is defined through the relationship with land. Land is therefore the repository of the family and community's (living, living dead, the yet-to-be-born and animate and inanimate citizens of the world) physical and spiritual connectedness. For Africans, land represents people's wellness, identity, integrity, security, economic vibrancy, purpose and meaning of life.

For indigenous African communities, land is more than a physical space. Land serves the good and wellbeing of community and humanity and does not belong to individuals. It is a communal treasure meant to take care of the people who work it for the betterment of the community. Land is *panarium,* it is life, and it provides food that must be shared by the community. Land is a vessel that holds family and community life together. In this sense, land represents human life and dignity, because the be-ing of family and community wellness is inextricably linked to land. The effects of dispossession in all its forms, including land, property and other material possessions, are devastating (Ngcukaitobi, 2018; Seleoane, 2014). We argue that dispossession, including that of land, results in loss of financial independence (Plaatje, 1991). It contributes to pathologies; psychological and emotional trauma, loss of identity, loss of meaning, feelings of insecurity and worthlessness, directionless, dependency and loss of human dignity. It is for this reason that Africans cannot define personhood, family and community in isolation from the land. Our view is that decolonised community psychology must place these post-colonial realities of Africans at the centre of the curriculum design. Community psychology should therefore advocate for land restitution and redistribution and foreground an understanding of personhood and community wellbeing in relation to land.

In this last experiential account by the second author, we demonstrate how personal narratives matter in our conception of epistemic justice that centres land, community, and African cosmology.

Personalising the Historical and Communal for Epistemic Justice

I, a black man (second author), was born in Lady Selbourne, Pretoria. My father worked at a factory in Pretoria and my mother was a domestic worker and stayed in the back yard of his uncle's home. Black people were later dispossessed of their land in Lady Selbourne. I am the second born in a family of six brothers and a sister. I am told that, at the time that I was born, my father proposed to marry my mother and paid *lobola*. As culture dictated at the time, having paid *lobola*, my grandparents requested my mother, their daughter in law (*makoti*) to relocate and join them in Nylstroom, now Modimolle (Limpopo Province) on Mr. Opperman's farm where they worked and resided. I was 2 years old when we went to stay with my grandparents. We have been informed that farmers who occupied this farm before it was sold to Mr. Opperman dispossessed this land from my great grandparents and turned them into labourers. My father continued to work in Pretoria and only came to visit us once every 3 months. This made him a migrant labourer.

Mr. Opperman's farm comprised only three families. These were, however, big households. Grandparents, parents and children from the three families constituted the workforce for the farm. My grandfather had ten children. In the African culture, uncles, aunts, grandparents, nieces, cousins and grandchildren constitute a family (Bujo, 1998). My grandfather built a house for each one of his children when they reached a certain age. In between these ten houses, he built three thatched roof rondavels. Two of them were used as kitchens, and one was mostly unoccupied as it was reserved for occasional visitors. Some of my aunts and uncles found work in nearby towns. The fortunate ones, like my father, went further away and found work in Pretoria. This is where he met my mother.

As young children growing up, we observed an interesting phenomenon in my grandfather's household. At regular intervals we would have three male strangers (Tumpu, Maphutha and Lelele) just dropping in and staying in the unoccupied rondavel. They would stay with us for a few weeks and in some instances, such as during harvest season; they would stay for 2 or 3 months. The three men were regarded as *balwetse ba tlhoganyo*, people with psychiatric conditions. We did not know where they came from, but our grandfather told us that God sent them, *ke bana ba Modimo*/God's children and that in the African culture they were part of our family and community. The two families that lived with us on the farm referred to them as *balwetse ba ga* Baloyi or psychiatric patients belonging to the Baloyi family. We all emulated our grandfather's spirit of *ubuntu*, and treated them as human beings, despite their psychiatric conditions. He taught us to see and treat people first of all, as human and spiritual beings. Upon reflection later in my adult life as an academic and trainer of clinical psychologists, what I learned from my grandfather was the manifestation of the meaning of *ubuntu/botho/* humanness (Nobles, Baloyi, & Sodi, 2016). Despite being a farm labourer, my grandfather had "his piece of land" where he could plant crops to feed his family. This also applied to the other three families on the farm. The communal approach to living on the farm, which included the three men that periodically visited my grandfathers' homestead, demonstrates *botho* in action. This is the Pan African Humanness, an African paradigm that espouses the richness of the life experiences, history and traditions of people of African ancestry (Nobles et al., 2016), that my grandfather taught the rest of the family.

When I first introduced the *African Epistemology* course in the mid-2000s in the training programme of clinical psychologists at the University of South Africa (UNISA), it was based on the above *ubuntu* philosophy and pedagogical value system. The course deconstructed Western psychology by placing working with indigenous people, the paradigms, theories, methodologies, cultural experiences and practices for pedagogic relevance and epistemic justice at the centre of the training of clinical psychologists. In my new university (Sefako Makgatho Health Sciences University), the African Epistemology course is a full 2-year course, comprising of theory and practice. Students in the clinical psychology course are introduced to African ethics, which sensitises them to epistemic and social justice issues. The core orientation of the African Epistemology course is to define African psychology and psychotherapy from the viewpoint of the African cultural experience. The starting point in working with people of African descent and understanding their humanity starts by introducing students to:

- The African conception of personhood, family and community or human beings—humanity and the human identity as defined by our humanness.
- How to negotiate entry into the client, family and community systems in a humane, respectful and polite manner—guided by the philosophy and ethic of *ubuntu.*
- How students locate themselves (their identity, relevance, position, power etc.) in families and communities they work with)
- The co-construction and theorisation of indigenous people's reality (psychological impact of poverty, health problems, unemployment, women, elderly and children abuse, landlessness, economic deprivation etc.) in communities projects using community connections/networks and local knowledge production paradigms.
- How to harmonise (intervene) using indigenous African communal approaches, theories, methodologies, and harmonisation modalities and the meanings for African families and communities emerging from these interactive processes.

My experience of introducing distinctively African centred pedagogy in psychology discourses was as a result of the realization by both black and white students and psychologists who noted and accepted the inadequacies of Western psychology when applied to indigenous clients and communities. Ethically, this constitutes an epistemological violation and injustice. Decolonisation of the psychology curriculum in general and community psychology empowers students to be culturally competent to work within the African socio-cultural frameworks in ways that are humane and respectful of people's relatedness.

Towards a Decolonised African-Centred Community Psychology

Through the experiential encounters outlined above, we point to some of the ways in which we are building decolonising technologies through an assemblage of resistances and forms of courage that harness alternative ways of being. These queer ways of being valorise disruption in order to register the presence and experience of Africans in their full diversity. For us, disruption means that we recognise that the people living and experiencing everyday life in the various contexts are experts in knowing about their own circumstances and lives. It is therefore imperative for our teaching and research to seriously consider this multiplicity of knowledges and allow them space in our classrooms. Decolonising is about thinking from within and without. It is not just a process of eradication—it is about getting to the relational norms of being. It requires us to ask questions about who we are as a people with multiple voices, and being given opportunities to respond appropriately and ethically to the demands of the world we live in. Nthabiseng Motsemme (2004, p. 5) argues:

> [M]aking efforts to link our stories to the everyday practices and fractured meanings of existence of individuals and communities is indeed a challenge, but also part of the unavoidable search of telling free and democratic stories.

Therefore, if we are to imagine decolonisation, we need to create the spaces for multiple converging and diverging stories to be told. It is at these often-uncomfortable intersections that the possibility for decolonisation may be initiated. Through our own experience, we have illustrated that within the colonial university is decolonial desire and praxis. Paperson (2017) calls those who dream and work for change, "decolonizing dreamers". These dreamers "wreck, scavenge, retool, and reassemble the colonizing university into decolonizing contraptions" (Paperson, 2017, p. xiii). We are decolonising dreamers and we teach to subvert coloniality and its binaries and essentialisms. We believe that binaries do not see our connection to land, spiritual identities, ever-present ancestors, and the unity of our bodies and minds.

We conclude by asserting that the philosophy of a relevant, meaningful and culture sensitive African psychology and community psychology curriculum in Africa must take the African conception of personhood, identity, family, community and land as its starting point. To understand and appreciate the relationship between one's identity, family and community, one has to adopt an epistemology and methodology that is congruent with an African worldview (Kovach, 2010b; Smith, 2012), premised on the philosophy of *ubuntu*. To understand African experiences using a foreign epistemology constitutes both an epistemic and an ethical injustice. Viewed from a de-colonial perspective, our respective personal accounts/stories exemplify one way of implementing/inaugurating an African-centred-pedagogy both at a theoretical and methodological levels, drawn from the oral tradition of knowledge production and learning. The current psychology and community psychology curriculum in South Africa require epistemological deconstruction and redesign to reflect reality from the viewpoint of the African experience, based on some of the proposals espoused in this chapter.

References

Ahmed, S. (2009). Embodying diversity: Problems and paradoxes for black feminists. *Race Ethnicity and Education, 12*(1), 41–52.

Andreotti, V. O., Stein, S., Ahenakew, C., & Hunt, D. (2015). Mapping interpretations of decolonization in the context of higher education. *Decolonization: Indigeneity, Education & Society, 4*(1), 21–40.

Anonymous. (1986). Some thoughts on a more relevant or indigenous counselling psychology in South Africa: Discovering the socio-political context of the oppressed. *Psychology in Society, 5*, 81–89.

Baloyi, L., & Ramose, M. B. (2016). Psychology and psychotherapy redefined from the viewpoint of the African experience. *Alternation Special Edition, 18*, 12–35.

Bujo, B. (1998). *The ethical dimension of community: The African model and the dialogue between north and south*. Nairobi, Kenya: Paulines Publications Africa.

Canham, H. (2014). Outsiders within: Non-conformity among four contemporary black female managers in South Africa. *Gender in Management: An International Journal, 29*(3), 148–170.

Crossman, P., and Devisch, R. (1999). Endogenisation and African Universities: Initiatives and Issues in the Quest for Plurality in the Human Sciences, Leuven, The Netherlands, Katholieke Universiteit Leaven.

de Sousa Santos, B. (2014). *Epistemologies of the south: Justice against epistemicide*. London New York: Routledge.

Duncan, N. (2013). *Community psychology: Analysis context and action*. Cape Town, South Africa: Juta.

Duncan, N., Bowman, B., Naidoo, T., Pillay, J., & Roos, V. (2008). *Community psychology in South Africa: Theory, context and practice*. Cape Town, South Africa: UCT Press.

Fanon, F. (1967). *The wretched of the earth* [1961], (Trans.). Constance Farrington. Harmondsworth: Penguin.

Freire, P. (1972). *Pedagogy of the oppressed*. Kääntänyt Myra Bergman Ramos, London: Penguin Books.

Freire, P. (1998). *Pedagogy of freedom: Ethics, democracy, and civic courage*. Lanham, MD: Rowman and Littlefield.

Fryer, D. (2008). Some questions about "The history of community psychology". *Journal of Community Psychology, 36*(5), 572–586.

Gordon, L. R. (2006). *Disciplinary decadence: Living thought in trying times*. Boulder, CO: Paradigm.

Grosfoguel, R., Oso, L., & Christou, A. (2015). 'Racism', intersectionality and migration studies: Framing some theoretical reflections. *Identities, 22*(6), 635–652.
Gyekye, K. (2002). Person and community in African thought. In P. H. Coetzee & A. P. J. Roux (Eds.), *Philosophy from Africa: A text with readings* (2nd ed., pp. 179–312). New York: Oxford University Press.
Kiguwa, P., & Canham, H. (2010). Creating a culture of thinking? Reflections on teaching an undergraduate critical social psychology course. *Acta Academica, 2010*(Suppl. 2), 62–90.
Kovach, M. (2010a). Conversation method in indigenous research. *First Peoples Child & Family Review, 5*(1), 40–48.
Kovach, M. (2010b). *Indigenous methodologies: Characteristics, conversations, and contexts.* Toronto, Canada: University of Toronto Press.
Lazarus, S. (1988). *The role of the psychologist in South African society: In search of an appropriate community psychology* (Unpublished Doctoral Dissertation). Cape Town, South Africa: University of Cape Town.
Lazarus, S., & Seedat, M. (1995). *Community psychology in South Africa.* Presented at the Fifth Biennial Conference of the Society for Community Research and Action, University of Illinois, Chicago, IL.
Makalela, L. (2017). Bilingualism in South Africa: Reconnecting with Ubuntu: Translanguaging. In O. Garcia, A. Lin, & S. May (Eds.), *Bilingual and multilingual education* (pp. 297–309). New York: Springer.
Maldonado-Torres, N. (2007). On the coloniality of being: Contributions to the development of a concept. *Cultural Studies, 21*(2–3), 240–270.
Manganyi, N. C. (1973). *Being-black-in-the-world.* Johannesburg: Raven Press.
Marx, J. (2004). Postcolonial literature and the Western literary canon. In N. Lazarus (Ed.), *The Cambridge companion to postcolonial literary studies* (pp. 83–96). Cambridge: Cambridge University Press.
Mignolo, W. D. (2009). Epistemic disobedience, independent thought and decolonial freedom. *Theory, Culture & Society, 26*(7–8), 159–181.
Mkhize, N. (2004). Psychology: An African perspective. In D. Hook, N. Mkhize, P. Kiguwa, & A. Collins (Eds.), *Critical psychology* (pp. 25–52). Lansdowne, South Africa: UCT Press.
Montero, M. (2007). The political psychology of liberation: From politics to ethics and back. *Political Psychology, 28*(5), 517–533.
Motsemme, N. (2004). The meanings in silence: Memory. *Rhodes Journalism Review, 2004*(24), 4–5.
Naidoo, A. V. (2000). *Inaugural Lecture. Community psychology: Constructing community, reconstructing psychology in South Africa.* Stellenbosch, South Africa: University of Stellenbosch.
Ngcukaitobi, T. (2018). *The land is ours: South Africa's first black lawyers and the birth of constitutionalism.* Johannesburg: Penguin Random House.
Nobles, W. W., Baloyi, L., & Sodi, T. (2016). Pan African humanness and Sakhu Djaer as praxis for indigenous knowledge systems. *Alternation Special Edition, 18*, 36–59.
Nwoye, A. (2018). African psychology and the emergence of the Madiban tradition. *Theory & Psychology, 28*(1), 38–64.
Nyamnjoh, F. B. (2005). A relevant education for African development—Some epistemological considerations. *Africa Development, 29*(1), 161–184.
Odora Hoppers, C. A. (2002). *Indigenous knowledge and the integration of knowledge systems: Towards a philosophy of articulation.* Claremont, South Africa: New Africa Books.
Painter, D., Terre Blanche, M. J., & Henderson, J. (2006). Critical psychology in South Africa: Histories, themes and prospects. *Annual Review of Critical Psychology, 5*, 212–235.
Paperson, I. (2017). *A third university is possible.* Minneapolis: University of Minnesota Press.
Pinar, W. F. (2012). *What is curriculum theory?* New York, NY: Routledge.
Plaatje, S. T. (1991). *Native life in South Africa before and since the European war and the Boer rebellion.* Athens, OH: Ohio University Press.

Pollock, G. (1999). *Differencing the canon: Feminist desire and the writing of art's histories*. Hove, England: Psychology Press.

Ramose, M. B. (2002). The philosophy of Ubuntu and Ubuntu as a philosophy. In P. H. Coetzee & A. P. J. Roux (Eds.), *Philosophy from Africa: A text with readings* (pp. 230–237). Cape Town: Oxford University Press.

Ramose, M. B. (2005). *African philosophy through Ubuntu*. Harare: Mond Books.

Ramose, M. B. (2016). Interview with professor Mogobe Ramose. *Journal on African Philosophy, 14*, 93–100.

Ramose, M. B., & Baloyi, L. (n.d.). *Questioning educational methodologies in pursuit of epistemic and social justice in South Africa*. (Unpublished manuscript).

Ramoupi, N. L. L. (2014). African research and scholarship: 20 years of lost opportunities to transform higher education in South Africa. *Ufahamu: A Journal of African Studies, 38*(1), 269–286.

Ratele, K. (2015, March 3). *Transformation of higher education should not be skin deep, UNISA*. Retrieved from http://www.unisa.ac.za/news/index.php/2015/03/transformation-of-higher-education-should-not-be-skin-deep/

Ratele, K. (2017). Editorial: Six theses on African psychology for the world. *Psychology in Society, 54*, 1–9.

Reyes Cruz, M. R., & Sonn, C. C. (2011). (De) colonizing culture in community psychology: Reflections from critical social science. *American Journal of Community Psychology, 47*(1–2), 203–214.

Seleoane, M. (2014). *Without fear or favour: Political essays*. Pretoria, South Africa: Ears Publishing.

Sharpe, C. (2016). *In the wake: On blackness and being*. Durham: Duke University Press.

Smith, L. T. (2012). *Decolonizing methodologies: Research and indigenous peoples* (2nd ed.). London/New York: Zed Books.

Snow, D. L., Grady, K., & Goyette-Ewing, M. (2000). A perspective on ethical issues in community psychology. In J. Rappaport & E. Seidman (Eds.), *Handbook of community psychology* (pp. 897–917). Boston, MA: Springer.

Teo, T. (2010). What is epistemological violence in the empirical social sciences? *Social and Personality Psychology Compass, 4*(5), 295–303.

Toro, P. A. (2005). Community psychology: Where do we go from here? *American Journal of Community Psychology, 35*(1–2), 9–16.

Tuck, E., & Yang, K. W. (2012). Decolonization is not a metaphor. *Decolonization: Indigeneity, Education & Society, 1*(1), 1–40.

wa Thiong'o, N. (1993). *Moving the centre: The struggle for cultural freedoms*. London/Portsmouth, NH: J. Currey/Heinemann.

Walsh, C. (2012). "Other" knowledges, "other" critiques: Reflections on the politics and practices of philosophy and decoloniality in the "other" America. *Transmodernity, 1*(3), 11–27.

Woodard, J. B., & Mastin, T. (2005). Black womanhood: Essence and its treatment of stereotypical images of black women. *Journal of Black Studies, 36*(2), 264–281.

One Peace for You, One Peace for Me: Pacification Versus Territorial Rights

Cristina J. Montiel and Erwine S. Dela Paz

Discourses about pacification first emerged during Spanish overseas conquests, when colonizers' rights to other peoples' property, dispossession, and free commerce were regarded as natural laws (Neocleous, 2010). Collective resistance against such decrees served as the source of Spain's presumed right to wage foreign wars and quell local resistance, in order to expand Spain's territory, power and economic influence (Neocleous, 2013). Philip II of Spain called these militarized foreign interventions as acts of pacification rather than conquest (Neocleous, 2010). Likewise, other European and Western nations such as the US extended their territory and carried out pacification drives in their respective colonies (Miller, 1982). Through the world's colonial history, pacification campaigns established economic control over a suppressed population, in exchange for so-called social order and protection from those who were recalcitrant to the law (Neocleous, 2013).

Such patterns of pacification may likewise arise in contemporary peace negotiations between central states and armed groups fighting for their independence. States may use peace talks as a discursive space where they negotiate to obtain cooperation from the armed groups. We point out that colonial narratives of pacification repeat themselves in modern-day peace talks, as central governments grapple with ethnic resistance movements.

Our paper argues that colonial storylines of pacification may be reproduced in former colonies, as sovereign central governments in the Global South subjugate their nation's peripheral ethnic groups. We suggest that a pacification narrative continues to be forwarded during intra-state peace talks, and may carry a storyline of covert suppression and economic exploitation from the center to the periphery. At the other side of the peace table, among marginalized ethnic groups, emerge decolonizing initiatives that demand territorial rights and economic sovereignty. We define

C. J. Montiel (✉) · E. S. Dela Paz
Department of Psychology, Ateneo de Manila University, Quezon City, Philippines
e-mail: cmontiel@ateneo.edu

G. Stevens, C. C. Sonn (eds.), *Decoloniality and Epistemic Justice in Contemporary Community Psychology*, Community Psychology,
https://doi.org/10.1007/978-3-030-72220-3_12

territorial rights as an exclusive space where political power can be exercised independently, while economic sovereignty as the independent power of a government to decide about its economy (Lowe, 1985; Stilz, 2009).

We present our argument by first reviewing the nature of US colonial pacification, especially in the Philippines. We elucidate on the colonialist aspect of pacification by discussing its destructiveness and constructiveness, and support of capitalism during colonization. We then forward the dual concepts of a *post-colonial pacification narrative* aimed to suppress periphery-based liberation forces vying for territorial sovereignty and a *decolonizing storyline* that asserts territorial and sovereignty rights. To expound on the emergence of pacification narratives between conflicting groups, the second part of our chapter looks at recent peace talks between Christians and Muslims in Mindanao, Southern Philippines.

Patterns of Pacification During the Colonial Period: A Review of US Colonialism and Philippine Anti-US Struggles

During the US colonization of the Philippines, the US pacified the Filipinos by representing themselves as foreign friends, rather than conquerors, and by ensuring that personal and religious rights of the Filipinos were protected (Miller, 1982). The US, not Filipinos, deployed the *pacification* lexical to describe a colonial relationship of plunder and domination. For example, in 1900 US President William McKinley declared in his presidential speech (McKinley, 1900):

> In my last annual message I dwelt at some length upon the condition of affairs in the Philippines. While seeking to impress upon you that the grave responsibility of the future government of those islands rests with the Congress of the United States, I abstained from recommending at that time a specific and final form of government for the territory actually held by the United States forces and in which as long as insurrection continues the military arm must necessarily be supreme. … I have steadfastly pursued the purpose so declared, employing the civil arm as well toward the accomplishment of **pacification** and the institution of local governments within the lines of authority and law.

Also at the turn of the twentieth century, US First Philippine Commission President J.G. Schurman (1900) applauded General Lawton's military ventures in the Philippines, in the name of pacification:

> Yet one cannot help feeling regret that our hero's work was so prematurely cut short, and that he has not been spared to fulfill the great duty which awaits us in the Philippines—the **pacifying** of the inhabitants, the binding up of their wounds, and the establishment under the stars and stripes of that golden age of liberty and justice for which the Filipinos have still to look forward.

The US also began to build a new social order by providing education, maintaining economic benefits and by enforcing law and order (Ileto, 2017). Despite such representations, Philippine independence struggles continued to resist US occupation and assert domestic sovereignty (Miller, 1982). We explain the nature of colonial pacification by looking at how the US pacified the Philippines. We characterize the nature of colonial pacification as destructive-constructive, and capitalistic (Davenport, 2004; McMichael, 2015; Neocleous, 2016).

The Destructive-Constructive Face of Colonialism

During the US colonial era, military and police powers were used to subdue resistance and pacify local peoples. For example, after the US gained control over the Philippines, they employed military offensives against locals, as armed battles took place when Filipinos continued to fight for their independence (Miller, 1982). Led by Emilio Aguinaldo, the Philippine revolutionary government resisted foreign rule by declaring war against the US. This led to the death of thousands of Filipino, and the destruction of numerous places. The war ended after Emilio Aguinaldo was captured, pledged allegiance to the US, and asked his followers to surrender (Miller, 1982). The US had used their military power to suppress and stop the insurrection in the name of pacification.

The US also enacted several colonial laws such as the Sedition Law, which made it illegal for Filipinos to advocate independence; and the Brigandage Law, which claimed to protect the Filipinos from bandits (Cullinane, 2009). But local revolutionaries were branded as brigands who must be hunted to maintain law and order.

Aside from subduing rebellions, colonial pacification was also characterized by building a new social order in controlled areas (Rigakos, 2011). For example, US president McKinley issued a policy called *Benevolent Assimilation*, which expressed that the personal and religious rights of the Filipinos would be protected by the US (Miller, 1982). The construction of a new social order included providing education which aimed to civilize and prevent Filipinos from taking arms against the US (Ileto, 2017; Maca, 2017; Wesling, 2011). US occupiers also established a civil government and a Philippine constabulary or police force (Cullinane, 2009). The US promised good governance to help Filipinos develop as a nation under the flagship of the colonizer (Ileto, 2017).

Pacification carries an economic storyline of material gain and wealth expansion (Schirmer & Shalom, 1987). With the new social order in place, the US then took control of Philippine economics and used its territory to gain access to China's market (Keliher, 2007). After pacifying the Philippines, the US acquired raw materials, and used Philippine territory as a military strong point. Furthermore, the US exerted economic control over the Philippines by enacting laws that favored US economic advantage (Mactal, 2014). For example, the Payne-Aldrich Act allowed importation of goods from the US into the Philippines without tariff and quotas, while subjecting some Philippine exports to the US with quotas (Ooi, 2007). Hence, with the US ensuring security in the Philippines through their police power, they then began gaining economic control and profits from the Philippines, a significant product of pacification's success (Neocleous, 2016).

Filipino Anti-US Struggles: Toward Economic and Territorial Sovereignty

The struggle of Filipinos to liberate their country from the US spanned 4 years (Ileto, 2017), resulting in the deaths of around 50,000 Filipinos and 4000 US soldiers (Tan, 2002). Filipinos leading the fight for independence included Emilio Aguinaldo with military generals like Antonio Luna and Gregorio Del Pilar (Dioso, 2004; Sheridan & Turot, 1998; Turot, 1981). Resistance against US occupation came in two forms—anti-US propaganda and armed confrontations with colonial forces.

Nationalistic manifestos encouraged Filipino revolutionaries to continue their fight for Philippine independence (Ileto, 2017). These political declarations exposed the sufferings of Filipinos under US rule, and emphasized the necessity of national liberation. For example, Aguinaldo fired up local anti-colonial struggles as he pleaded to his fellow revolutionaries, at the historical juncture after colonial ownership of the Philippines had been transferred from Spain to the US (Garcia, 1969; Turot, 1981):

> Filipinas! Beloved daughter of the ardent sun of the tropics… Thou hast longed for independence, and thine emancipation from Spain has come…But thou, Filipinas, flower of the ocean, delicate flower of the East, still weak, scare eight months weaned from thy mother's breast, has dared to brave a great and powerful nation such as is the United States, with thy little army barely disciplined and shaped. Ah, beloved brethren, all this is true; and still we say we will be slaves to none, nor let ourselves be duped by gentle words.

Aside from using manifestos as means to express decolonization, armed conflicts took place in different regions of the country (Barnett, 2005). In Manila, Filipino soldiers under Aguinaldo fought against US combat troops in towns like Caloocan and Pasig (De Viana, 2006). In the town of Calumpit in Luzon, a fierce battle took place between the US troops and Filipino soldiers headed by General Antonio Luna who attempted to stop the advance of the US toward the town of Pampanga where the Philippine republic retreated after its defeat in Manila (Keenan, 2001). In the town of Samar in Visayas, Filipino soldiers killed 44 US soldiers. The US then retaliated killing hundreds of Filipino soldiers and civilians (Ileto, 2017; Linn, 2000). In addition to confronting the US in the open battlefield, Filipinos employed guerilla warfare, engaging in sabotage and other forms of harassment in population centers (Ileto, 2017). Massive secret guerrilla resistance increased in towns occupied by the US as Filipinos continued to struggle for their sovereignty.

Colonial Pacification and Sovereignty Storylines

Let us now unpack more deeply the clashing meanings of pacification and sovereignty during the colonial period. We see one identical US-Philippines inter-nation relationship, framed by two chasmically divergent political narratives. To understand the impact of such colonial storylines on the lives of a colonized people, we analyze how nations and peoples are positioned toward each other by the narratives constructed by states about inter-nation relationships.

A positioning lens provides useful tools for analyzing political storylines in global conflict, by dissecting how participants talk and write about a conflict (Moghaddam, Harre, & Lee, 2008a). From a positioning viewpoint, what is important is how people construct meanings about a conflict, rather than how a conflict is structured or resolved (Moghaddam, Harre, & Lee, 2008b). Interestingly, "…the meanings one group of actors assigns to what is going on may be very different from the meanings assigned by other people involved in the very same episode" (Moghaddam et al., 2008b, p. 10).

Briefly, positioning analysis posits that meanings of the words used by actors perform a specific social task (Slocum-Bradley, 2010). We see identities as categories and attributes attached to a social group, whereas positions refer to rights and duties. Narratives present the storyline of a group's discourse. Positioning theory asserts that discursive productions evoke identities and positions based on the storyline (Davies & Harre, 2007; Slocum-Bradley, 2010; Van Langenhove & Harre, 1994). A positioning lens illuminates the psychological understanding of (de)colonization during peace talks, by unpacking how intergroup utterances allocate rights and duties to group identities included in the political conversation.

To elucidate on how storylines, identities, and rights and duties are differentially constructed by two conflicting parties, let us apply positioning analysis to the US colonial era. Table 1 applies a positioning lens to understand two views of US

Table 1 Meanings of peace during the US colonial period in the Philippines: pacification vs. territorial sovereignty

Two meanings of peace during the US colonial period	Socially constructed national identities	Rights and duties
The US pacification storyline	*US* • Maintains security • Creator of social order through education	*US* • Has the right to exploit economy • Has the duty to ensure safety of colony
	Philippines • Uneducated • Incapable of running a government	*Philippines* • Has the right to education • Has the duty to follow US governance
The Philippine storyline about territorial sovereignty	*US* • Exploiters of natural resources • Colonizers	*US* • Has the right to give Philippines its independence • Has the duty to honor their promise of independence
	Philippines • A revolutionary government • An independent nation	*Philippines* • Has the right to be independent • Has the duty to fight for independence

pacification drives—what pacification meant to the colonizer, and what it meant to the colonized. Filipinos fought for national sovereignty, against these so-called pacification drives. Table 1 presents these bifurcated meanings.

In the second column of Table 1, national identities are specified, indicating how such conflicting identities of nations emerge from the clashing narratives of pacification and sovereignty. For example, from the pacification storyline, the US was a creator of social order, and Filipinos were uneducated and incapable of self-governance. But from a sovereignty narrative, the US was an exploiter of natural resources, while Filipinos were revolutionaries forming a national government. Table 1 also shows how differently each of the two storylines allocate rights and duties to each nation. The pacification narrative of the colonizer claims that the US has the right to exploit the Philippine economy, while Filipinos are expected to comply with US governance. On the other hand, the sovereignty narrative demands that the US grant the Philippines independence, and that Filipinos hold the right to fight for independence.

Our analysis demonstrates how, through a simple political play of linguistic acrobatics, "Different meanings may be given to the same action …The antithetical meanings come from discordant storylines, sometimes on the scale of cosmic narratives" (Moghaddam et al., 2008b, p. 18). By employing positioning analysis on the colonial narratives of US pacification drives and Philippine sovereignty struggles, we illustrate how a single historical colonial relationship can take on two clashing meanings, with massive implications especially on the lives of the colonized.

Pacification Vs. Territorial Sovereignty Storylines in Domestic Peace Negotiations: An Empirical Study of Peace Talks Between the Philippine Government and the Moro Islamic Liberation Front

But colonial narratives do not die easily. Similar pacification-versus-sovereignty contestations persist until today, between central governments and local ethnic groups fighting for their liberation. In contemporary Philippine history, a parallel conflictual pattern of pacification versus territorial rights conflagrates in the Mindanao-based conflict. Muslims from Mindanao in Southern Philippines continue to press for their own Moro independence and defy what they perceive as unjust annexation by the post-colonial and Christian–led Philippine Government (Abubakar, 2004). Armed resistance is waged through various Moro liberation fronts, and push the Philippine Government to the peace negotiating table with the Moros. However, as of this writing, peace remains elusive as negotiations after negotiations have broken down (Abubakar, 1989, 2004; Lagarde, 2015).

The first two peace efforts took place in Tripoli and in Jeddah where the Moro National Liberation Front agreed to self-govern several regions in Southern Philippines. Both peace agreements failed due to disagreements over the number of provinces to be included in the pact, and whether to conduct a plebiscite (Abubakar, 2004; Bacani, 2005). The third peace agreement focused on merging the armed forces on both sides of the conflict in order to foster economic growth in Mindanao. This agreement integrated members of the Moro National Liberation Front into the Philippine Armed Forces and Philippine National Police (Abubakar, 2004; Plank, 2015).

Subsequently, other Moros split from the Moro National Liberation Front and formed the Moro Islamic Liberation Front, citing ideological and strategic differences. This new Moro Front expanded their fighting forces and established parallel governments throughout Mindanao (Adam, 2018). The last two peace agreements, referred to as the Memorandum of Agreement on Ancestral Domain, and the Bangsamoro Basic Law, granted the Moro Islamic Liberation Front a Bangsamoro homeland (Rodil, 2009). However, both peace-making efforts collapsed due to so-called constitutionality issues expressed by the Philippine Supreme Court and the Philippine Congress respectively (Bacus, 2015; Candelaria, 2009).

During peace talks with the Moro liberation fronts, the Philippine Government repeatedly stressed constitutionality, ceasefires, and disarmament. On the other side of the peace talks, the Moros demanded territorial autonomy and pursued the creation of a Moro government. We argue that the clashing narratives of pacification and territorial sovereignty thrived not only during US colonial times, but also in contemporary clashes between the Philippine government and the marginalized Moro groups.

The similarities across historical moments are startling. In both instances, the dominant yet seemingly predatory groups—the colonizing US and the contemporary Christian-led Philippine government—silence resistance groups in the euphemistic name of pacification, cessation of direct violence, and calls for disarmament. The dominant stances of the US over the Philippines during the colonization era, and of the Philippine government over the MILF, present similar coloniality patterns of establishing control over indigenous peoples. Similarly, in both historical moments, the subjugated populations' collective narrative is one of struggle for national or territorial sovereignty. We now look at the Bangsamoro peace talks between the Philippine Government and the MILF.

Examining Pacification Narratives during Contemporary Peace Negotiations

Peace negotiations between Philippine Government and MILF include the passage of the Bangsamoro Basic Law. In this second part of our paper, we examine speeches by both the Philippine Government and the Moro Islamic Liberation Front during such peace negotiations. We worked with a data corpus consisting of speeches during the peace negotiations between the Government of the Philippines and the Moro Islamic Liberation Front (MILF) over 5 years, from 2011 to 2016.

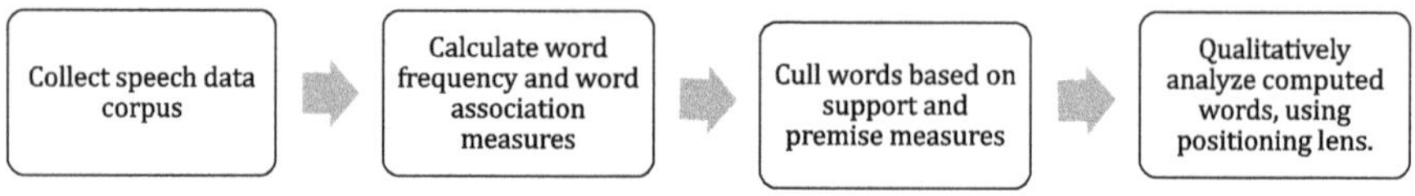

Fig. 1 Data collection and analysis of speeches by the Philippine government and the Moro Islamic Liberation Front during peace negotiations (2011–2016)

We selected speeches made only in reference to the Bangsamoro peace talks. Such speeches included demands, appreciative comments, and agreements between both parties. From the Philippine Government's side of the conflict, we collected 56 speeches by key state actors such as the Philippine president and government peace negotiators. From the other side of the peace table, we acquired 36 speeches by the MILF peace representatives and the head of their liberation front.

We asked whether the Christian-dominated government forwarded a post-colonial agenda of pacification and integration. More specifically, we investigated whether the Philippine Government's language of peace is described through the cessation of direct, armed violence and the formulation of peace agreements within the bounds of the Philippine constitution. We likewise examined whether the MILF resistance group forwarded a decolonizing narrative of territorial sovereignty. Such a storyline would be embodied in calls for self-determination and the need to rectify historical injustices committed against the Bangsamoro people.

To answer our questions, we first implemented social computing, counting and associating words often used in the speeches. We then conducted qualitative analysis of our computed results, in order to elucidate deeper meanings of pacification and resistance talks. Figure 1 capsulizes our research method.

Quantitative data analysis through social computing. We used the software RapidMiner to carry out text analysis. This software quantitizes qualitative data by using algorithms or a series of computer programing instructions that carry out specific tasks in a given data pool (North, 2012). To examine our collection of speeches, we employed algorithms that determined which words were often used in both Philippine Government and MILF data speeches, then compared frequency patterns across data sets (North, 2012).

Subsequently, we employed an algorithm that identified sets of words associated with each frequently used word (North, 2012; Tan, Steinbach, & Kumar, 2006). Such associated words are called premises (North, 2012). To illustrate the concept of premise, let us refer to Table 2, which summarizes our computational results. Table 2 shows that in the speeches made by the MILF, the words *Bangsamoro* and *government* premise the frequently-used word *peace*. This implies that there is a high chance that when the words *Bangsamoro* and *government* are uttered by the MILF, the word *peace* would also be found in the same speech text.

We also computed for the percentage of speeches that contained both the frequently used words and their premises. This measure is called support (Tan et al., 2006). For example, Table 2 indicates that 69.6% of the total speeches of the

Table 2 Premises of frequent words used in speeches by Philippine Government and Moro Islamic Liberation Front

Philippine government			Moro Islamic Liberation Front		
Premise	Frequent words	Support	Premise	Frequent words	Support
Milf, government, bangsamoro, members	Peace	0.696	Milf, thank, government, good, bangsamoro	Peace	0.718
Peace, government, bangsamoro, members	Milf	0.696	Peace, thank, government, good, bangsamoro	Milf	0.718
Peace, milf, bangsamoro, agreement, members	Government	0.643	Peace, milf, government, good, bangsamoro	Thank	0.718
Peace, milf, government, agreement, members	Bangsamoro	0.643	Peace, milf, thank, good, bangsamoro	Government	0.718
Peace, milf, government, bangsamoro, members	Agreement	0.643	Peace, milf, thank, government, bangsamoro	Good	0.718
Peace, milf, government, bangsamoro, agreement	Members	0.643	Peace, milf, thank, government, good	Agreement	0.769

Philippine Government hold a premise word like *milf, government, bangsamoro,* and/or *members*, which appear(s) in the same speech as the frequent word, *peace*.

Interestingly, both the Philippine Government and the Moro Islamic Liberation Front use almost identical words during their peace talks with each other. Computational results in Table 2 reveal that during the peace process, both groups frequently used: *peace, MILF, government,* and *agreement*. The term *bangsamoro* is often uttered by the Philippine Government, and likewise precedes the frequently-used words of the Moro Front.

We note, however, that only the MILF discourse is punctuated by the words *good* and *thank you*. One way to understand the meaning of the MILF's *good* and *thank you* utterances is by examining its social effects through positioning analysis. We surmise that such types of utterances reveal a kind of ingratiating orientation by a marginalized group, where negotiated rights are not claimed with a sense of entitlement, but are instead accepted with good gratitude toward a high-power group. At a later part of our chapter, we point out that although both groups tended to use the same words as they talked peace with each other, these identical words took on different meanings, when analyzed in their linguistic context.

Qualitative data analysis through positioning theory. We likewise employed positioning analysis to examine peace-related storylines in post-colonial conditions, as the Philippine Government negotiated with the Moro Islamic Liberation Front. We returned to the mathematically-derived results of our speech analysis, consisting of word frequency counts and word association measures. We then shifted to qualitative analysis by drawing on the analytical power of a positioning lens.

We nuanced the meaning of these contextualized words, by using the frequently-used words and their premises (see Table 2) as search keys in the speeches. We then read and re-read the original full text to get a sense of what the words meant in their actual linguistic context. We then used the frequently-uttered words, their identified premises, and their linguistic contexts in the speeches, to draw out the colonizing and decolonizing storylines of both the Philippine Government and MILF. We completed our positioning analysis of the speeches on both sides of the negotiating table, by examining how both the Philippine Government and MILF discursively constructed their identities, and defined their collective rights and duties.

Our positioning findings show data patterns of pacification and decolonization from speeches made by Philippine Government and Moro Islamic Liberation Front respectively. Qualitative analysis also reveals a Philippine Government pacification storyline that continues to assert its sovereignty over the MILF, whereas the MILF's decolonizing narrative advances a sub-state political identity that governs itself. Table 3 below presents our qualitatively derived storylines, that show how each party positioned the other through their use of linguistically-contextualized words.

In the second and third columns of Table 3, we find discursive-positioning accounts of how each storyline characterizes the respective identities, and sets of rights and duties, of the two parties involved in the peace talks. In the first storyline, the Philippine Government identifies itself as responsible for maintaining security and creation of social order whereas the MILF is identified as a minority Filipino group. Hence, the Philippine Government claims the right to negotiate with the MILF and a duty to maintain territorial unity (i.e., no secessions), while the MILF is allocated the right to participate in a meaningful autonomous governance, and a duty to unite with the Philippine government. The second storyline contains a decolonizing narrative from the Moros. The MILF views the creation of a sub-state as a way of achieving peace in the region. This suggests that the Moros aim to distinguish themselves as different from the Philippines. The Moros assert that a "good" Bangsamoro Basic Law (BBL) from the Philippine Congress will facilitate the peace process. They also thank the government for their support of the Comprehensive Agreement and recognize the Philippine Government as the legitimate government in Manila.

The decolonizing storyline identifies the Philippine government as a central government committed to the peace agreement, and the MILF as a sub-state with a distinct political identity. Such creation of a sub-state emphasizes the MILF's demands for territorial sovereignty from the central government in Manila. In this second storyline, the Philippine Government holds the right to support the BBL and the duty to ensure passage of the BBL in the Philippine Congress. On the other hand, MILF claims their right to self-governance and determination, with the duty to cooperate with the Philippine Government.

Table 3 Meanings of peace during contemporary peace talks in the Philippines: pacification vs. territorial sovereignty

Two meanings of peace during negotiations between the Philippine government and the Moro Islamic Liberation Front (MILF)	Socially constructed national identities	Rights and duties
The Philippine Government's pacification and integration storyline We ask the ***MILF*** to withdraw its armed ***members*** from these barangays and coordinate—Through our AHJAG (ad hoc joint group overview)—So we could do the right thing in the right way. The comprehensive **agreement** on the ***Bangsamoro*** acknowledges a different narrative of our national being, one that would bridge our majority minority divide toward a shared future where fellow Filipinos live in ***peace*** under one flag in an undivided territory. This ***Agreement*** will enable MILF combatants and other disenfranchised segments of the Moro population to participate in meaningful autonomous governance. Greater participation of the regional ***government*** in the management of mineral resources; and a host of other items on education and economic reforms.	*Philippine government* Maintains security Creator of social order by sharing a common future	*Philippine government* Has the right to negotiate with MILF Has the duty to keep the territory undivided
	MILF A minority group Filipino	*MILF* Has the right to participate in meaningful autonomous governance Has the duty to unite with Philippine government
The MILF's territorial sovereignty storyline For the ***MILF***, we view the state-substate formulation as a formula for ***peace*** and unity in this country. ***Thank*** you for the numerous declarations and commitment of support to the comprehensive ***Agreement*** on the Bangsamoro. It is to the interest of the ***government*** in Manila that a viable and progressive Moro entity is in place in Mindanao. There are other challenges ahead of us, but I am sure the moment a ***good*** BBL comes out of congress, everything gradually falls into place and becomes easy.	*Philippine government* Committed to BBL Central government	*Philippine government* Has the right to support BBL Has the duty to ensure passage of BBL in congress
	MILF Sub-state Political entity	*MILF* Has the right to self-govern and determination Has the duty to cooperate with the Philippine gov't

Conclusion

In this chapter, we argue that peace-related narratives have been employed by more dominant groups to silence struggles for territorial sovereignty. At the height of global colonial conquests, US militarily controlled and muted its colonized peoples, euphemistically labeling its predatory overseas missions as pacification drives. In contemporary post-colonial societies, a similar peace-related storyline may be claimed by dominant central governments, as they engage in peace talks with break-away ethnic groups. For example, in its peace discourse with the Moro Islamic Liberation Front (MILF), the Philippine Government forwards issues like the disarmament of MILF combatants, and the management of Mindanao mineral resources.

But subjugating pacification storylines continue to be countermanded by decolonizing and liberating narratives of local peoples' struggles for territorial sovereignty. During colonial times, local freedom fighters confronted foreign pacification forces in the open battlefield, during guerrilla warfare, and through propaganda movements. A similar decolonizing discourse arises during modern-day struggles for territorial sovereignty by marginalized ethnic groups. For example, the MILF has waged decades of armed struggles against the Philippine Government, based on political demands for an autonomous Moro territory. Our study's analysis of MILF speeches during recent peace talks shows how the Moro liberation group claims its right to self-govern while simultaneously acknowledging its duty to cooperate with the central government.

Our results forward cautionary notes about using peace-related lexicals as linguistic expressions, during intergroup conflict management in formerly colonized countries. The word *peace* may have acquired a subjugating connotation, associated with historical memories of colonial pacification drives, and post-colonial efforts to mute local resistance efforts against external domination. Our study shows how dual concepts of a post-colonial pacification narrative by a central government, and a decolonizing storyline by peripheral or ethnic groups, are expressed during intrastate peace negotiations.

We propose a politically-sensitive nuancing of the meaning of *peace* at negotiating tables, when asymmetrically-related groups face each other as they discuss their armed conflicts. On one side of the peace conversation sits a more dominant group attuned to pacification storylines. On the other side stands a marginalized yet resistant group, vying for territorial sovereignty.

Parties engaged in peace negotiations may appear to have a common goal of achieving peace. Clashing groups may even use the same words in their intergroup peace conversation. Further, during domestic peace negotiations, both groups may bear mirror images of coloniality. On one hand, the powerful central government forwards pacification demands such as disarmament or ceasefires. On the other hand, resistance groups at the peace table may offer ingratiating narratives, mirroring compliant orientations toward a modern-day colonizer.

But on a deeper level, the linguistic terrain of peace negotiations—and historical collective actions as well—may reveal powerfully divergent storylines of pacification and territorial sovereignty. We caution that attending solely to pacification narratives during contemporary peace talks may gradually re-subjugate indigenous peoples, contribute to the breakdown of peace talks, and bring about the emergence of new liberation factions which continue to demand territorial sovereignty.

References

Abubakar, C. A. (1989). Moro ethno-nationalist movement. In K. David & S. Kadirgamar (Eds.), *Ethnicity, identity, conflict and crisis* (pp. 108–125). Hong Kong: Arena Press.

Abubakar, C. A. (2004). Review of the Mindanao peace processes. *Inter-Asia Cultural Studies, 5*(3), 450–464. https://doi.org/10.1080/1464937042000288732

Adam, J. (2018). Bringing grievances back in: Towards an alternative understanding of the rise of the Moro Islamic Liberation Front in the Philippines. *Journal of the Humanities and Social Sciences of Southeast Asia And Oceania, 174*(1), 1–23.

Bacani, B. R. (2005). *The Mindanao peace talks: Another opportunity to resolve the Moro conflict in the Philippines/Benedicto R. Bacani*. Washington, DC: USIP.

Bacus, R. J. (2015). Autonomy or secession: Analysis of the constitutionality of the Bangsamoro basic law. *Ateneo Law Journal, 59*(4), 1433–1456.

Barnett, G. (2005). A trilogy of wars: The Philippine revolutionary wars of 1896-97, the Spanish-American war in the Philippines in 1898, and the Filipino-American war, 1899-1902. *Military History, 2*, 66.

Candelaria, S. M. (2009). Postscript to the supreme court MOA-AD judgment: No other way but to move forward. *Ateneo Law Journal, 54*(1), 269–273.

Cullinane, M. (2009). Bringing in the brigands the politics of pacification in the colonial Philippines, 1902-1907. *Philippine Studies, 57*(1), 49.

Davenport, C. (2004). The promise of democratic pacification: An empirical assessment. *International Studies Quarterly, 3*, 539.

Davies, B., & Harre, R. (2007). Positioning: The discursive production of selves. *Athenea Digital, 12*, 242–259.

De Viana, A. V. (2006). *The I-stories: The Philippine revolution and the Filipino-American war as told by its eyewitnesses and participants*. Manila: University of Santo Tomas Pub. House.

Dioso, M. M. (2004). *A trilogy of wars: The Philippine revolutionary wars of 1896-7, the Spanish-American war in the Philippines in 1898, and the Filipino-American war, 1899-1902*. Pittsburgh, PA: Dorrance Pub.

Garcia, M. (1969). *Aguinaldo in retrospect: A volume issued to commemorate the centenary of General Emilio Aguinaldo y Famy: With documents on the Philippine-American war, 1898-1901, and the first Philippine Republic*. Manila: Philippine Historical Association.

Ileto, R. C. (2017). *Knowledge and pacification: On the U.S. conquest and the writing of Philippine history*. Quezon City: Ateneo de Manila University Press.

Keenan, J. (2001). *Encyclopedia of the Spanish-American & Philippine-American wars*. Santa Barbara, CA: ABC-CLIO.

Keliher, M. (2007). Anglo-American rivalry and the origins of U.S. China policy. *Diplomatic History, 31*(2), 227–257. https://doi.org/10.1111/j.1467-7709.2007.00613.x

Lagarde, M. D. (2015). Clearing the pathways to peace: Law, policy, and the Mindanao conflict. *Ateneo Law Journal, 59*(4), 1141–1204.

Linn, B. M. (2000). *The Philippine war, 1899–1902*. Lawrence: University Press of Kansas.

Lowe, A. (1985). The problems of extraterritorial jurisdiction: Economic sovereignty and the search for a solution. *International and Comparative Law Quarterly, 4*, 724.

Maca, M. (2017). American colonial education policy and Filipino labour migration to the US (1900–1935). *Asia Pacific Journal of Education, 37*(3), 310–328.

Mactal, R. B. (2014). Ang epekto ng digmaang Pilipino-Amerikano sa kabuhayan at antas ng pamumuhay sa Maynila, 1898-1901. *Malay, 26*(2), 36–52.

McKinley, W. (1900, December 3). *Presidential speech* [written]. Retrieved from https://genius.com/amp/William-mckinley-presidential-speech-1900-written-annotated

McMichael, C. (2015). Urban pacification and 'Blitzes' in contemporary Johannesburg. *Antipode, 47*(5), 1261–1278. https://doi.org/10.1111/anti.12160

Miller, S. C. (1982). *Benevolent assimilation: The American conquest of the Philippines, 1899–1903*. New Haven: Yale University Press.

Moghaddam, F. M., Harre, R., & Lee, N. (2008a). *Global conflict resolution through positioning analysis*. New York: Springer.
Moghaddam, F. M., Harre, R., & Lee, N. (2008b). Positioning and conflict: An introduction. In F. M. Moghaddam, R. Harre, & N. Lee (Eds.), *Global conflict resolution through positioning analysis* (pp. 3–20). New York: Springer.
Neocleous, M. (2010). War as peace, peace as pacification. *Radical Philosophy, 159*, 8–17.
Neocleous, M. (2013). The dream of pacification: Accumulation, class war, and the hunt. *The Social Studies, 9*(2), 7.
Neocleous, M. (2016). The logic of pacification: War-police accumulation. *Athenea Digital, 16*(1), 9–22. https://doi.org/10.5565/rev/athenea.1738
North, M. (2012). *Data mining for the masses*. Global Text Project, c2012.
Ooi, K. G. (2007). *Southeast Asia: A historical encyclopedia, from Angkor Wat to East Timor*. Santa Barbara, CA: ABC-CLIO.
Plank, F. (2015). Not enough pieces of the cake? The Moro National Liberation Front (MNLF) in the Mindanao final agreement. *Asian Security, 11*(2), 154. https://doi.org/10.1080/14799855.2015.1071795
Rigakos, G. (2011). "To extend the scope of productive labour": Pacification as police project. In M. Neoclous & G. Rigakos (Eds.), *Anti-security* (pp. 57–84). Ottawa: Red Quill Books.
Rodil, R. B. (2009). Conflict between the Moro Liberation Fronts (MIM-MNLF-MILF) and the government in historical perspective. *Ateneo Law Journal, 54*(1), 256–262.
Schirmer, D. B., & Shalom, S. R. (1987). *The Philippines reader: A history of colonialism, neocolonialism, dictatorship, and resistance*. Quezon City: Ken.
Schurman, J. G. (1900). General Lawton as warrior, statesman, and man: Elegant tributes from President Schurman of the Philippine Commission, and the Rev. Peter MacQueen, a hero whom the nation mourns. *Leslie's Weekly*, 10–11.
Sheridan, R. B., & Turot, H. (1998). *Eyewitness accounts in 1900 of the Philippine revolution, the Filipino-American war and the U.S. occupation of the Philippines*. Manila: Trademark Publishing Corporation.
Slocum-Bradley, N. (2010). The positioning diamond: A trans-disciplinary framework for discourse analysis. *Journal for the Theory of Social Behaviour, 40*(1), 79–107. https://doi.org/10.1111/j.1468-5914.2009.00418.x
Stilz, A. (2009). Why do states have territorial rights? *International Theory, 1*(2), 185–213. https://doi.org/10.1017/S1752971909000104
Tan, S. K. (2002). *The Filipino-American war, 1899–1913* (p. c2002). Quezon City: University of the Philippines Press.
Tan, P. N., Steinbach, M., & Kumar, V. (2006). *Introduction to data mining*. Boston: Pearson Addison Wesley.
Turot, H. (1981). *Emilio Aguinaldo: First Filipino president, 1898–1901*. Manila: Foreign Service Institute, University of the Philippines.
Van Langenhove, L., & Harre, R. (1994). Cultural stereotypes and positioning theory. *Journal for the Theory of Social Behaviour, 24*(4), 359–372.
Wesling, M. (2011). *Empires proxy: American literature and U.S. imperialism in the Philippines*.

Index

G. Stevens, C. Sonn (eds.), *Decoloniality and Epistemic Justice in Contemporary Community Psychology*, Community Psychology,
https://doi.org/10.1007/978-3-030-72220-3

GPSR Compliance
The European Union's (EU) General Product Safety Regulation (GPSR) is a set of rules that requires consumer products to be safe and our obligations to ensure this.

If you have any concerns about our products, you can contact us on

ProductSafety@springernature.com

In case Publisher is established outside the EU, the EU authorized representative is:

Springer Nature Customer Service Center GmbH
Europaplatz 3
69115 Heidelberg, Germany

www.ingramcontent.com/pod-product-compliance
Ingram Content Group UK Ltd.
Pitfield, Milton Keynes, MK11 3LW, UK
UKHW022212051026
15624UKWH00001B/121
9783030722197